Recycling of Rare Earths

David J. Fisher

Published by **Materials Research Forum LLC**
Millersville, PA 17551, USA

Published as part of the book series
Materials Research Foundations
Volume 119 (2022)
ISSN 2471-8890 (Print)
ISSN 2471-8904 (Online)

Print ISBN 978-1-64490-178-6
ePDF ISBN 978-1-64490-179-3

Distributed worldwide by

Materials Research Forum LLC
105 Springdale Lane
Millersville, PA 17551
USA
http://www.mrforum.com

Printed in the United States of America
10 9 8 7 6 5 4 3 2 1

Table of Contents

Materials Research Forum LLC
https://doi.org/10.21741/9781644901793

Introduction

There was a distant time when any chemistry teacher would preface his introduction to the rare earths by pointing out that this was a misnomer: they are far from rare (making up to 240mg/g of the Earth's crust, table 1) when compared with more familiar elements such as gold and platinum; sometimes being many thousands of times more abundant. Cerium is almost as common as copper. And the teachers who pointed this out did so at a time when just about everybody was carrying around rare earths in their pockets, in the form of mischmetal 'lighter flints'. It was never their abundance which made them rare, but rather the absence of rich localized deposits; a deficit which of course still plagues geologists and miners.

But rare-earth elements are now critically important to the present-day world, due largely to their magnetic and phosphorescence properties. Their ubiquity in the communications and green-energy fields has indeed made them strategically important materials, on a par with uranium, lithium, cobalt or helium. Given that the mining of ore and the separation of the pure elements is expensive, especially when the deposits are in geographically or geopolitically inconvenient locations, recycling of any material which is already to hand becomes very attractive. Given also that many of the items which contain rare-earth elements are subject to the fickleness of consumers, there is a literal wealth of discarded products which already constitutes an environmental problem. The potentially 'virtuous circle' of reducing junk-mountains while simultaneously reducing strategic vulnerability then becomes a matter of finding economically viable processes for the recycling of rare-earth elements.

The rare-earth elements conventionally comprise 15 lanthanides (cerium, dysprosium, erbium, europium, gadolinium, holmium, lanthanum, lutetium, neodymium, praseodymium, promethium, samarium, terbium, thulium, ytterbium) plus yttrium and scandium; the latter two being included because of their similar chemical properties to those of rare earths and their consequent tendency to be found in the same ores. Many of their names reflect the fact that they were first discovered by Swedish workers. Not all of the rare earths are of interest in the present context however: promethium, in particular, possesses no stable or long-lived isotopes and is thus more likely to be synthesized anew than to be recycled.

They are generally trivalent, although cerium can be tetravalent and europium divalent. Atomic radii normally increase with increasing atomic number, but the opposite occurs in the case of the rare earths: this is the so-called lanthanide contraction, and occurs because electrons in the f-orbitals do not screen other electrons from the attraction of the nucleus.

Materials Research Forum LLC
https://doi.org/10.21741/9781644901793

This in turn causes the rare earths to exhibit diverse chemical reactivities. They are often split into two sub-groups: light and heavy. The elements between lanthanum and europium are termed light rare-earth elements, while those between gadolinium and lutetium are termed heavy rare-earth elements. Scandium is not included in either sub-group, and the placement of yttrium is also rather anomalous. Both types of rare earth are found in the same ore deposits, but the light elements (137.8ppm) are more abundant than the heavy ones (31.3ppm) in the Earth's crust. They can also be divided into two groups on the basis of their tendency to leach into the environment from, for example, computer mother-boards. Samarium, europium, gadolinium and dysprosium have a greater tendency to contaminate than do the remainder of the rare earths.

Table 1. Average abundances of rare-earth elements in the Earth's crust
Data from Lide, D.R., Handbook of Physics and Chemistry, CRC Press, 1997

Rare Earth	Abundance (mg/g)
cerium	66.5
dysprosium	5.2
erbium	3.5
europium	2
gadolinium	6.2
holmium	1.3
lanthanum	39
lutetium	0.8
neodymium	41.5
praseodymium	9.2
samarium	7.05
scandium	22
terbium	1.2
thulium	0.52
ytterbium	3.2
yttrium	33

The rare earths are now everywhere: in electronic display screens for flat-screen televisions, computer-monitors and cell-phones, in silicon chips, rechargeable batteries, lenses, light-emitting diodes and fluorescent lamps. Neodymium is now commonly associated with high-strength magnets, even though holmium in fact possesses the highest magnetic strength of any element, and such magnets make possible the construction of miniature drones and compact electric motors in general. They are present in automobile catalytic converters as well as in electrically driven vehicles and wind turbines; thus making them an essential component of devices which will help to reduce global warming. Lanthanum and cerium are used in catalysis in quantities which rival that of other rare earths in magnets. Europium, terbium and yttrium are essential colour-producing phosphors in video screens.

The down-side is that many of these current applications, being subject to rapid innovation and fashion (table 2), lead to mountains of junked consumer products in the form of so-called 'e-waste'. At the same time, these junk-mountains can well contain increasingly scarce or difficult-to-source elements such as dysprosium, erbium and europium. The average abundances listed in table 1 hide the fact that the rare earths are not uniformly distributed in the Earth's crust. The entire history of civilization has been played out against the backdrop provided by the fortuitous presence of useful resources in various countries, thus leading to colonization, invasion and wars as well as mutually profitable trade. The uneven distribution of the 132 million tonnes of rare-earth sources (tables 3 to 5) - and thus supply - threatens to become another example of this tendency, and effective recycling promises to alleviate its ill effects to some extent.

Table 2. Average lifetime of electronic products
Data from Computer Technology Association, 2014

Product	Year of Appearance	Lifetime (years)
desk-top computer	1980	5.9
lap-top computer	1980	5.5
cell-phone	1985	4.7
flat-screen television	1990	7.4
DVD devices	1997	6.0
digital camera	1998	6.5
smart-phone	2000	4.6
tablet computer	2008	5.1

Table 3. Sources of rare-earth elements

Ore	Rare Earth
Allanite	lanthanum, yttrium
Apatite	lanthanum
Bastnaesite	lanthanum, yttrium
Eudialyte	lanthanum
Fergusonite	lanthanum, yttrium
Iimoriite	yttrium
Kainosite	lanthanum, yttrium
Loparite	yttrium
Monazite	lanthanum, thorium
Mosandrite	lanthanum
Parisite	lanthanum
Pyrochlore	lanthanum
Rinkite	lanthanum
Steenstrupine	lanthanum, thorium
Synchysite	lanthanum
Xenotime	yttrium
Zircon	lanthanum

Table 4. World reserves of rare-earth oxides
Data from US Geological Survey, 2018

Country	Share of Total (%)
China	33.33
Brazil	16.67
Vietnam	16.67
Russia	13.64
India	5.23

Australia	2.56
Greenland	1.14
USA	1.06
South Africa	0.65
Canada	0.63
Malawi	0.11
Malaysia	0.02

Table 5. Rare-earth production by country

Country	Fraction of Total Production (%)
China	78.7
Australia	15
Russia	2.2
Brazil	1.5
Thailand	1.2
India	1.1
Malaysia	0.2
Vietnam	0.07

Because of their reactivity, they occur together in numerous silicate, carbonate, oxide and phosphate ores and these in turn are found in particular geological milieux such as carbonatites, alkaline igneous systems, ion-adsorption clay deposits and monazite-xenotime bearing placer deposits. In the Earth's sedimentary system, the parent-daughter radioactive element pair lutetium/hafnium is strongly fractionated[1] relative to samarium/neodymium, and this is attributed to the high resistance to chemical weathering which is exhibited by the zircon, $(Zr,Hf)SiO_4$. Zircon-bearing sandy sediments which are located on, and near, the continents have very low lutetium/hafnium ratios while deep-sea clays have up to three times the chondritic lutetium/hafnium ratio. Turbulent currents carry sandy material, with low lutetium/hafnium ratios, onto the ocean floor. The above-listed bastnaesite, monazite and loparite – together with lateritic clays – are major sources, but it is unusual to find rare earths in economically viable ore deposits. The very definition of the word 'ore' has always been an essentially economic one. It has even been suggested that the ocean bottom deposits may be a far more promising source than those on land. The latter deposits are largely in the form of alkaline igneous rocks such as pegmatites and carbonatites, heavy mineral placers and constituents of coal. In some places, Pacific mud deposits are said to contain up to 1000 times more rare-earth elements than do the known land reserves. Such muds also require minimal processing: e.g. leaching. Extraction of the rare earths usually involves dissolving the ore in acidic or alkaline solutions and heating the ore to up to 500C.

Carbonatite-associated rare-earth deposits are the most important source of the elements, but their origin is somewhat unclear[2]. The deposits are located along the western margin of the Yangtze Craton, where Cenozoic strike-slip faults were related to Indo-Asian continental collision. The Cenozoic carbonatites tend to be very rich in barium, strontium and rare earths, with $^{87}Sr/^{86}Sr$ ratios greater than 0.7055. The carbonatites were probably formed by melting of the sub-continental lithospheric mantle which had been previously metasomatized by high-flux rare-earth element- and CO_2-rich fluids originating from subducted marine sediments. It is suggested that cratonic edges, particularly along convergent margins, are optimum configurations for the generation of giant rare-earth deposits.

The above tables conceal the fact that, in spite of its relatively meagre share of deposits, the US previously dominated the rare-earth market. The move of China towards a more capitalist economy and its new involvement in high-technology products, especially wind turbines of late, has completely up-ended the market. With this advantage, and with a burgeoning self-sufficient domestic market, China understandably had an interest in restricting rare-earth exports. When China limited its rare earth exports in 2010 it produced a huge world-wide spike in rare-earth prices. The price of cerium, for example, rose from some 300$/kg early in 2011 to a maximum of 11,000$/kg in mid-2012 before starting to fall equally quickly when China increased rare-earth production and eased export restrictions. Illegal rare-earth trading in China began to increase markedly in 2007 and this was provoked mainly by growing restrictions on the legitimate trade. This illegal production forced the overall prices to fall. At the end of 2016, prices again began to increase and China now seems to be quite determined to use rare earths as a weapon in international trade and a bargaining chip in international diplomacy.

One response to this has been to seek alternatives to the rare earths by, for example, using special copper alloys or iron/nickel composites to make magnets, instead of neodymium. Fluorescent lights and light-emitting diodes can be made in such a way as to avoid the use of terbium, europium and ytterbium. Light-emitting diode lamps are typically found[3] to comprise 8 and 14wt% of inorganics, 34 and 36wt% of polymers and 50 and 58wt% of metals when leached using *aqua regia* and nitric acid, respectively.

Cheaper magnetic materials can certainly be found for use in electric cars and wind turbines, but it is unclear whether they can ever rival the device miniaturization which is made possible by the rare earths.

The increasing gap between rare-earth demand and supply is naturally driving a search for alternative resources, with even the waste product known as coal fly ash being considered as a possible source because it can contain more rare earths than does an ore. This is an exciting prospect as it suggests the possibility of 'mining' (and remediating) the detritus of the environmentally unpopular fossil-fuel industry for useful materials.

One potential counterbalance to the Chinese advantage in rare-earth sourcing is Greenland. It seems that the rare-earth deposits there may be vast, although there is considerable opposition to their exploitation, on environmental grounds.

Rare-earth element distribution patterns were long-ago determined[4] for samples of Amîtsoq gneiss from the Godthåb and Isua areas of Greenland. All had fractionated rare-earth patterns with light rare-earth enrichment although some leucocratic components exhibited extreme depletion in heavy rare-earths. Anorthosites from Ameralik and Buksefjord exhibited light rare-earth enrichment, with undetectable or small europium anomalies.

The contents of rare-earth elements (La, Ce, Nd, Sm, Eu, Gd, Tb, Tm, Yb, Lu), as well as of tantalum, thorium and hafnium in kimberlites and inclusions from Greenland were deduced[5] by means of neutron activation. All of the samples had highly fractionated rare-earth distribution patterns. The La/Yb ratios in the Greenland kimberlites (hypabyssal facies) varied from 111.8 to 188.4, while the total rare-earth contents ranged from 204.8 to 380.3ppm. No europium anomaly was found.

Rare-earth abundances were determined[6], by activation analysis, in samples from the Fiskenaesset complex. The rare-earth data for rocks and plagioclases were distinct from those of many other anorthositic complexes, and the abundances were among the lowest detected for plagioclase. The plagioclases exhibited a positive europium anomaly of about 10, together with a depletion of heavy rare-earths relative to light ones. Mafic separates were enriched in heavy rare-earths, relative to light ones, and generally exhibited no europium anomaly.

Rare-earth element data for 14 gabbro and syenogabbro samples from the Tugtutôq younger giant dyke complex, and for 3 anorthosite inclusions, supported other evidence that the anorthosites were early fractionates that formed at depth from a magma which

Materials Research Forum LLC
https://doi.org/10.21741/9781644901793

was similar to the liquid that produced the giant dykes themselves[7]. All of the samples were enriched in light rare-earths, and the absolute rare-earth concentrations increased with increasing degree of differentiation in chilled marginal samples and in cumulates. Marked positive europium anomalies in the early cumulates decreased with continued fractionation, and negative europium anomalies could be present in the most differentiated cumulates. The absence of appreciable europium anomalies in chilled samples indicated that large quantities of anorthosite, and thus large positive europium anomalies, could not have formed directly from the younger giant dyke complex magma. The late fractionates of the younger giant dyke complex and the early augite syenite phase of Ilimaussaq exhibited very similar rare-earth patterns. The later agpaites of Ilimaussaq were greatly enriched in rare earths and exhibited marked negative europium anomalies.

Rare-earth concentrations were determined[8] for circa 2800 million year old Nûk gneisses from the Buksefjorden region of south-west Greenland. The samples comprised dioritic to granodioritic gneisses and synplutonic mafic dykes. The early Nûk gneisses, diorites and tonalites have slightly fractionated rare-earth patterns; supposed to result from the partial melting of garnet-bearing amphibolite or granulite. Early Nûk trondhjemitic gneisses possess downward-convex patterns with clear positive europium anomalies. Most of the later Nûk grey gneisses have extremely fractionated linear patterns. The rare-earth patterns which are found in the late Nûk Ilivertalik granite complex are slightly fractionated, but with a high overall abundance. Two sets of synplutonic mafic dykes had strongly fractionated patterns.

Cumulus apatite, sphene, feldspar, amphibole and biotite from the pulaskite of the Kangerdlugssuaq alkaline intrusion have been analyzed[9] for rare-earth elements. The apatite was particularly rich in rare earths, containing 3.6% of cerium, and exhibited a steep chondrite-normalized pattern, enriched in light rare-earths. The other minerals exhibited an enrichment in light rare earths, but the sphene had a cerium peak on a chondrite-normalized plot. The rare-earth partition-coefficient values showed that the light rare earths were preferentially accommodated by apatite rather than sphene. The differences in the coefficients in turn resulted from differences in the coordination of the rare earths in the two minerals.

In a deposit which was formed by the incomplete re-equilibration of ultrabasic and quartzofeldspathic reactants, transport of rare earths over several metres occurred during diffusion-controlled metasomatism of the protoliths[10]. The larger resultant concentration range of europium was greater than 2 orders of magnitude. The rare-earth content generally increased towards the marginal zones. The rare earths were the least mobile species in the aqueous transporting medium and the final rare-earth distribution was suggested to be governed mainly by the crystal structures of talc, tremolite, hornblende and chlorite.

The rare-earth patterns in 3.8 gigayear-old Isua iron formations are generally flat, although samples with a positive, negative or zero europium anomaly were to be found[11]. Diagenesis and metamorphism did not appreciably change the rare-earth patterns[12]. The petrology and rare-earth geochemistry of clastic rocks from the 3800 million-year belt indicated[13] that the rare-earth patterns for muscovite-biotite gneisses are strongly fractionated, with variable europium anomalies. Garnet-biotite schists have less fractionated light rare earths and exhibit a slope reversal for heavy rare earths. These represent a mixed felsic-mafic protolith.

Perovskite-group minerals, ABX_3, from the intrusive ultramafic alkaline Gardiner Complex of East Greenland, range from almost pure $CaTiO_3$ to cerium-based loparite[14]. Chemical zoning in the perovskites is controlled by the substitutions,

$$2Ca^{2+} \rightarrow Na^+ + \Re^{3+}$$

on the A-site and

$$2Ti^{4+} \rightarrow Fe^{3+} + Nb^{5+}$$

on the B-site. Other trace elements include thorium, strontium, aluminium, silicon, zirconium, tantalum and tin. Chondrite-normalized rare-earth patterns exhibit an enrichment in light rare earths in the case of perovskite, loparite, apatite, melilite and diopside. The mean perovskite/apatite partition coefficients for 4 Gardiner rocks were: La = 10.4, Ce = 13.8, Nd = 13.9, Sm = 9.9, Eu = 7.7, Gd = 5.2, Tb = 5.6, Tm = 5.5, Yb = 2.7 and Lu = 1.6. This indicated that the perovskite concentrated all of the rare earths to a much greater extent than did apatite. Enrichment of light rare earths occurs in both perovskite and apatite.

Nepheline syenites form part of the Gardar Province of south Greenland, where fractionation of the syenitic magmas has resulted in a peralkaline residual lujavritic magma containing rare-earth elements together with yttrium, zirconium and niobium[15]. Syenites were affected by a metasomatism that was associated with the intrusion and evolution of younger syenites. Differing fluids evolved from each new syenite and tended to produce 2 compositionally distinct products. The metasomatism which was associated with one of these fluids resulted in an extensive redistribution of rare earths and related elements; to be found in apatite, titanite and fluorcarbonate minerals. In the case of apatite, zoning involved the exchange,

$$Ca^{2+} + P^{5+} \rightleftharpoons \mathfrak{R}^{3+} + Si^{4+}$$

while the variation was less systematic in eudialyte and titanite but involved calcium, sodium, silicon, rare earths, yttrium, zirconium and niobium. In contrast to apatites from the nearby Ilímaussaq intrusion, there is no significant sodium present in the structures and exchange reactions involving Na^+, $\mathfrak{R}^{3+}$ and Ca^{2+} have not occurred[16].

Mesozoic undersaturated lamprophyre dykes from southern West Greenland exhibit rare-earth concentration levels which are comparable to those in similar rock-types found elsewhere[17]. A secular trend has been established[18] in Precambrian iron-formations, in that the oldest iron-formations exhibit only slight fractionation of light rare-earths relative to heavy rare-earths, and this fractionation increases with decreasing age of the iron formations.

Rare-earth data on discrete growth phases of the complex polyphase zircons from early Archaean Amîtsoq gneisses in Godthåbsfjord, south-west Greenland reveal[19], in Matsuda diagrams, steep positive slopes going from lanthanum to lutetium, a marked positive cerium anomaly and negative europium anomalies. These are consistent with growth in a melt. Non-cathodoluminescent zircon and late Archaean prismatic tip overgrowths instead exhibit flatter light rare-earth patterns and exhibit little or no europium anomaly.

The increasing demand for lanthanides has driven the investigation of stream sediment data on Greenland. The results for lanthanum, europium and ytterbium are considered to representative of the 7 rare earths which are sampled[20]. The spatial distributions and anomalies (La > 364ppm, Eu > 7.4ppm, Yb > 16ppm) show that certain provinces are enriched in one, two or three of these elements. Sedimental anomalous rare-earth samples from South Greenland, arising from rocks which host rare-earth deposits (carbonatite, peralkaline syenite, alkaline granite, non-alkaline granite) show that the rare-earth fractionation in each of these rock associations is distinct. When applied to data from the whole of Greenland, these differences make it possible to identify those samples which probably came from the same associations. Samples which arise from different lithologies also occupy differing regions in plots of La-Yb, Yb-Eu and La/Yb-Eu/Yb variation. The stream-sediment rare-earth properties additionally reflect those of the

Materials Research Forum LLC

https://doi.org/10.21741/9781644901793

source rocks because rare earths are preferably found in minerals that can survive in a stream-sediment environment.

The rare mineral, Moskvinite-(Y) $Na_2K(Y,\mathfrak{R})Si_6O_{15}$, has been discovered[21] at Kvanefjeld, as a replacement mineral which is associated with a mineral belonging to the britholite group. The empirical formula, based upon 15 oxygen atoms, is,

$$Na_{1.94}K_{0.99}(Y_{0.94}Yb_{0.03}Er_{0.03}Dy_{0.03}Ho_{0.01}Gd_{0.01})\square_{1.05}Si_{5.98}O_{15}$$

This coexistence of almost pure yttrium and a light rare-earth element mineral is explained in terms of the fractionation of rare earths and yttrium during the replacement of a prior rare-earth mineral. Theoretical calculations of the replacement of feldspathoids by natrolite showed that the resultant fluid would have a pH greater than 8; thus inhibiting large-scale mobility of the rare earths.

Another factor favouring Greenland is that, although the rare-earth resources in China are huge, most of them are associated with natural radionuclides. The perceived risk is that these radionuclides will inevitably pollute the environment, near and far, as Chinese rare earths inevitably spread in parallel with the export of their consumer electronics. The initial smelting of rare earths also poses serious environmental pollution problems. An investigation[22] of the migration of radioactive elements during the smelting of rare earths has shown that some 90% of the radioactive elements concentrate in residues while the remainder is distributed as waste water and gas. The radioactive waste residue generated during rare-earth production has thus created serious environmental problems. It has been suggested[23] that residue-leaching and thorium-separation should be applied to the proper disposal of radioactive waste residue from ion-absorbed rare earth separation industries. Factors such as type of leaching acid, acid concentration and liquid/solid ratio were considered and a multi-step leaching process was proposed. Following multi-step leaching with hydrochloric acid, the total leaching efficiencies of rare earths and thorium were greater than 98.14% and 99.07%, respectively. An extractant, sec-octylphenoxy acetic acid, was then used to separate thorium and enrich rare earths from the leachate residue. The separation efficiency of thorium was better than 99.53% and the lanthanide concentration in the concentrated solution was 223.19g/l. The recycling of LiCl-KCl eutectic salt wastes which contained radioactive rare-earth oxychlorides or oxides has been studied[24] with the aim of minimizing radioactive waste by using vacuum distillation. Vaporization of the LiCl-KCl eutectic was effective above 900C, under 5Torr, and condensation of the vaporized salt depended mainly upon the temperature gradient.

Materials Research Forum LLC
https://doi.org/10.21741/9781644901793

Recycling in a closed loop was able to ensure a high efficiency of recovery, with more than 99wt% being extracted.

At the same time however, the Chinese home market is suddenly becoming greedier for rare earths[25]. This is because concerns over anthropogenic climate change, and perhaps the present uncomfortable role of China as a major fossil-fuel polluter, are driving an enthusiasm for wind turbines; with their associated need for magnets. It is predicted that the annual demand for the relevant non rare-earth metals will be some 12 times higher in 2050 than in 2018; with the cumulative demand being up to 23 times larger. The cumulative copper and nickel needs for wind power correspond to up to 45.9% of Chinese copper reserves and up to 101%[*sic*] of Chinese nickel reserves. In the case of rare earths, a greater than 18-fold increase is expected up to 2050, with the cumulative demands for neodymium and dysprosium being up to 3.3% and up to 2.8% of domestic reserves, respectively. Recycling will thus play a critical role after 2050, even in China. A similar scenario is foreseen for Japan[26]. A flow analysis of neodymium focused on the waste flow of the final product and demand for, and waste of, neodymium up to 2050 were forecast by multivariate analysis. This showed that domestic wastage of neodymium was expected to increase from 3866 to 4217tons/year by 2050. Material recycling of neodymium might cause an additional increase in production.

Environmental effects

The dumping of tens of millions of tonnes of e-waste every year is meanwhile leading to the release of rare earths into the environment and their toxic effects upon human beings are little known. Rare earths in fact constitute a relatively new group of substances considered to be pollutants, and this novelty is due to explosion of their use in high technology. A recent study[27] of 45 elements in the population of Andalusia was based upon plasma samples[28] from 419 participants. Arsenic, copper, lead, selenium, antimony, strontium and bismuth were detected in more than 98% of the subjects, with the median levels of arsenic, mercury and lead being 1.49, 1.46 and 5.86ng/ml, respectively, and with a correlation existing between age and concentration in the case of arsenic, mercury, antimony and strontium. The total of elements was lower in those younger than 45 years old, suggesting the successful control of notorious pollutants in more recent times. There was a positive correlation between body-mass-index and the concentrations of barium, cerium, osmium, tin and ytterbium. Among the rare earths which were monitored, 7 of them were detected in more than 90% of cases. Yttrium and cerium were present in the highest concentrations, with median values of 0.19 and 0.16ng/ml, respectively. The plasma levels of 6 of the rare earths were higher among males, with a positive correlation found between concentration and age. In another study[29], inhabitants of the Canary

Materials Research Forum LLC
https://doi.org/10.21741/9781644901793

Islands were compared with those of Morocco; the point being that, although relatively close, there are significant differences in socio-economic development, lifestyle and underlying geology. The blood concentrations of 47 elements, including rare earths, were measured in similar cohorts in Morocco and in the Canaries; 124 and 120, respectively. The levels of iron, selenium, zinc, arsenic, cadmium, strontium and especially lead were much higher in Moroccan residents and this was attributed to the intensive mining in that country. There were also much higher total levels of rare earths in the Moroccan residents, but this was attributed to the careless handling of e-waste. In the case of the Canary Island inhabitants, there were higher levels of manganese. That was attributed to a more widespread use of motor vehicles. The concentrations of 48 elements, including rare earths, were determined[30] in whole-blood samples from 63 sub-Saharan immigrants who were suffering from anaemia, and the results were compared with those for a 78-strong control group. The levels of iron, chromium, copper, manganese, molybdenum and selenium were much higher in the control group than in the anaemic group; suggesting nutritional deficiency. The levels of silver, arsenic, barium, bismuth, cerium, europium, erbium, gallium, lanthanum, niobium, neodymium, lead, praseodymium, samarium, tin, tantalum, thorium, thallium, uranium and vanadium were higher in the anemic cohort, with a generally inverse dependence upon haemoglobin concentration. There was sometimes an inverse dependence upon blood iron level, suggesting that a higher rate of intestinal up-take was related to an iron nutritional deficiency. On the other hand, the higher lead and rare-earth levels in anaemic subjects were independent of the iron levels and suggested that those elements might be implicated in anaemia. A study was also made[31] of concentrations of rare earths and hormones in plasma from the inhabitants of an e-waste area and of a control area in Taizhou (China). The concentrations of lanthanum and cerium were much higher in the exposed group than in the control group. Only one thyroid hormone content was significantly higher in the exposed group, and this was associated with high lanthanum and cerium levels. Biomarkers of oxidative stress were also much higher in the exposed group, and such stress was behind effects of the rare earths upon the hormone. The internal exposure of rare earths in the inhabitants of the e-waste area was appreciable, and the compositional profile of those elements in the exposed group was different, due to e-waste treatment.

Metal levels were measured[32] in the blood and urine of e-waste recyclers at Agbogbloshie (Ghana) and compared with those of a control group. Samples from 100 e-waste recyclers and 51 controls were analyzed for 17 elements (Ag, As, Ba, Cd, Ce, Cr, Eu, La, Mn, Nd, Ni, Pb, Rb, Sr, Tb, Tl, Y). The mean blood levels of lead, strontium, thallium, and the urine levels of lead, europium, lanthanum, terbium and thallium were significantly higher

Materials Research Forum LLC
https://doi.org/10.21741/9781644901793

for recyclers than for the controls. Collectors and sorters tended to have higher elemental levels than other workers.

Other studies[33] had investigated the inhabitants of regions of rare-earth mining, including the association of tobacco smoking and indoor smoke with increased levels of rare earths and respiratory-tract damage. There is also rare-earth bio-accumulation in head-hair, and defective gene expression. As in the historic case of asbestos exposure, certain occupations increase contact with rare earths: e-waste processing, diesel-engine repair and cinematic film operation. The general population can also be exposed due to exhaust microparticulates arising from the CeO_2 in catalytic converters. Such diesel microparticulates have been studied, revealing pathological effects in animals. With regard to smoking, it was noted that tobacco plants and vegetables used in the manufacture of cigarette papers could accumulate elements from the environment. Some 32 elements which are related to high-technology use were detected[34] in tobacco, cigarette papers and filters. Significant differences in the levels of most of the elements were found in the tobacco and in the cigarette papers, and black tobacco cigarettes contained the highest levels. The paper which was used in home-made cigarettes could significantly concentrate the elements. Fast-burning, bleached and flavored papers also created higher levels of pollutants. Theoretical exposures could differ by up to 40 times, depending upon whether the cigarette was factory- or hand-made. From the point-of-view of anthropogenic sources, transfer mechanisms and the environmental behaviour of rare earths, lanthanum, gadolinium, cerium and europium are the most studied, revealing that the main anthropogenic sources include medical facilities, petroleum refining, mining, high technology, fertilizers, animal feed and e-waste recycling[35]. The rare earths are then spread through the environment via hydrological and wind-driven processes. The ecotoxicological effects include stunted plants, genotoxicity and neurotoxicity in animals, trophic bio-accumulation and toxicity in soil organisms. Human exposure occurs via the ingestion of contaminated water, food and air. Rare earths may cause nephrogenic systemic fibrosis in humans and severe damage to nephrological systems.

They also have an effect of course upon aquatic ecosystems, and creatures such as mussels: *mytilus galloprovincialis*, exposed to increasing dysprosium concentrations (0, 0.1, 1, 10mg/l) over a 28 day period[36] (18C, salinity = 30) indicated that dysprosium was responsible for a metabolic increase that was associated with glycogen expenditure, the activation of anti-oxidant, cellular damage and a loss of redox balance. This could seriously affect physiological functions, reproduction and growth. Lanthanum exposure[37] elicited a biochemical response in the mussels, as reflected by a lowered metabolism and the activation of anti-oxidant defences and biotransformation enzymes; particularly at intermediate concentrations. In spite of oxidative stress, resultant damage was avoided.

Enzyme inhibition demonstrated the neurotoxicity of lanthanum in this species, and it is regarded as being a threat to marine organisms. Standardized toxicity tests have been performed[38] on the oyster, *Crassostrea gigas*, in order to assess the effects of lanthanum and yttrium. The former possesses a greater toxicity than the latter, with median effect-dose concentrations of 6.7 to 36.1µg/l over 24 and 48h for lanthanum and 147 to 221.9µg/l over 24 and 48h for yttrium. The higher toxicity of lanthanum was attributed to its higher bio-availability in free ionic form. Lanthanum can be placed among the compounds most toxic to this oyster, while yttrium is of intermediate toxicity. As a final example, model nitrifying organisms, *Nitrosomonas europaea* and *Nitrobacter winogradskyi*, have been exposed[39] to simulated waste-waters which contained various levels of yttrium or europium (10, 50, 100ppm) and the extractant, tributyl phosphate (0.1g/l). Yttrium and europium additions of 50 and 100ppm inhibited *Nitrosomonas europaea*, even when hardly any of the rare earths were soluble. The presence of tributyl phosphate and europium increased Nitrosomonas *europaea* inhibition while tributyl phosphate alone did not appreciably alter the activity. In the case of *Nitrobacter winogradskyi* cultures, additions of europium or yttrium at any tested level led to marked inhibition, and nitrification ceased entirely upon tributyl phosphate addition. Model calculations revealed a strong pH-dependence of the rare-earth solubility.

Removal of these elements can only benefit the environment, as well as providing a large fraction of the rare earths required by industry. In addition to recuperating rare earths, processing of the waste would also recover gold and platinum-group metals. The recycling of 100,000 Apple iPhones can yield 1900kg of aluminium, 770kg of cobalt, 710kg of copper, 93kg of tungsten, 42kg of tin, 11kg of rare earths, 7.5kg of silver, 1.8kg of tantalum, 0.97kg of gold and 0.1kg of palladium. This possibility is important because recycling of the rare earths themselves is not easy, and simultaneous recuperation of gold for example could make the recycling of rare earths more economically viable even before recycling perhaps becomes a strategic necessity.

The separation of the rare-earth elements is difficult, which is why they were available only as so-called mischmetal for many years. Mischmetal continues to be useful because of its hydrogen-storage abilities, and finds a place in rechargeable nickel metal hydride batteries. New separation techniques have nevertheless had to be developed. These include the use of novel organic compounds which form complexes with the rare earths, and the subsequent exploitation of the differences in the solubilities of those complexes. Another possible method is to use *Gluconobacter* bacteria which produce acids from sugar and dissolve the rare-earth elements.

During bioleaching with *Gluconobacter oxydans* the gluconate, which mediates metal leaching, can be oxidized to 2-ketogluconate and 5-ketogluconate. At a pH of 6.0 or 9.0, without pH control, complexolysis was the predominant leaching mechanism and a higher rare-earth leaching efficiency was possible with gluconate, while 5-ketogluconate offered more efficient base-metal leaching[40]. At a pH level of 3.0, acidolysis predominated and the base-metal and rare-earth leaching yields were higher than those at any other pH levels. The highest base-metal and rare-earth leaching yields were observed by using gluconate at a pH level of 3.0 and were 100.0% for manganese, 90.3% for iron, 89.5% for cobalt, 58.5% for nickel, 24.0% for copper, 29.3% for zinc and 56.1% for rare earths. Heterotrophic bioleaching of rare earths and base metals from spent nickel-metal-hydride batteries has been investigated[41] with regard to the effect of phosphorus, in the form of $Ca_3(PO_4)_2$, KH_2PO_4 and K_2HPO_4, on *Gluconobacter oxydans* and *Streptomyces pilosus*. The source of the phosphorus affected the microbial acid production and thus the metal leaching. The use of K_2HPO_4 led to the highest organic acid production by both bacteria, and increasing the K_2HPO_4 concentration from 2.7 to 27mM increased pyruvic acid production by *Streptomyces pilosus* from 2.2 to 10.7mM. On the other hand, no metal was leached from the batteries when using *Streptomyces pilosus* and 1-step or 2-step bioleaching. In the case of *Gluconobacter oxydans*, the highest gluconic acid concentration (45.0mM) was produced at the lowest of the above K_2HPO_4 concentrations. Two-step bioleaching with *Gluconobacter oxydans* offered higher leaching efficiencies for iron, cobalt and nickel, while rare earths were better leached by using spent-medium bioleaching. Base-metal leaching was faster than that of rare earths when using either bioleaching method. It was concluded that a surplus of phosphorus should be avoided in bioleaching cultures when dealing with rare earths.

Table 6. Rare earths recoverable from consumer products

Product	Recoverable Rare Earths
Fluorescent lamps	cerium, europium, gadolinium, lanthanum, terbium, yttrium
LEDs	cerium, europium, gadolinium, lanthanum, terbium, yttrium
Plasma screens	cerium, europium, gadolinium, lanthanum, terbium, yttrium
Cathode-ray tubes	europium, yttrium
Permanent magnets	dysprosium, neodymium, praseodymium, terbium
Automobiles	dysprosium, neodymium, terbium
Mobile phones	dysprosium, neodymium, praseodymium, terbium
Hard disk drives	dysprosium, neodymium, praseodymium, terbium
Computers	dysprosium, neodymium, praseodymium, terbium
Household appliances	dysprosium, neodymium, praseodymium, terbium

The recovery of rare earths from scrapped products (table 6) will certainly become a viable industry if efficient means can be found for a) first separating them from other materials such as plastics and then b) separating them from each other and from other metals. Recycling can be closed-loop or open-loop. In the former, the recovered rare earths are re-used for similar purposes. In the latter, the recovered rare earths are used for other purposes. In that context, it is interesting to compare the above table with one showing the demands on rare earths made by various applications (table 7).

Table 7. Rare earths required by various applications

Application	Principal Rare Earths
magnets	69.4% neodymium, 23.4% praseodymium, 5% dysprosium, 2% gadolinium
batteries	50% lanthanum, 33.4% cerium, 10% neodymium, 3.3% samarium
auto-catalysis	90% cerium, 5% lanthanum, 3% neodymium, 2% praseodymium
oil-refining	90% lanthanum, 10% cerium
polishing	65% cerium, 31.5% lanthanum, 3.5% praseodymium
glasses	66% cerium, 24% lanthanum, 3% neodymium, 1% praseodymium
phosphors	69.2% yttrium, 11% cerium, 8.5% lanthanum, 4.9% europium
ceramics	53% yttrium, 17% lanthanum, 12% cerium, 12% neodymium

The first problem above is made all the more difficult because product-designers rejoice in making improvements which increase the durability of those products. Their embedded systems, laminated components and printed-circuit boards create recycling puzzles. A classic principle of conventional mining is the comminution rule which says that an ore should first be ground to a particle size which is half of that of the smallest of the particles of the material being mined. The constant trend toward ever-increasing miniaturization obviously militates against that rule. Printed-circuit boards and embedded systems increase the durability of components but reduce their size. Structurally integrated materials clearly make disassembly and recovery more difficult as the components may be screwed, bolted, glued or soldered together and have to be separated by crushing, grinding or shearing. Protective coatings of polymers, which improve anti-moisture protection, also have to be removed by dissolution or heat-treatment.

Materials Research Forum LLC
https://doi.org/10.21741/9781644901793

An examination of end-of-life electrical and electronic scrap was conducted[42] with particular regard to 16 mundane elements, 2 precious metals and 15 rare earths. The first group included copper (23% found in laptops), aluminium (6% found in computers), lead (15% found in DVD players) and barium (7% found in televisions). The second group included gold (316g/ton) in mobile phones and silver (636g/ton) in laptops. Most of the waste printed-circuit boards contained considerable quantities of rare earths, with scandium amounting to 31g/ton and cerium amounting to 13g/ton. Recycling of e-waste in the search for rare earths can be made more economically attractive by the presence of all of the other mundane but useful metals. Metal scrap which contains ferrous and non-ferrous alloys can be partially recovered, but not all metals can be effectively recovered. Depending upon the percentages present, iron and aluminium are obvious targets for recovery and recycling. As shown above, e-waste also typically contains quite large amounts of copper, silver, gold and palladium and these can be economically recovered. Their recovery generally involves pyrometallurgical or hydrometallurgical processing. The former involves melting, and consequently high energy requirements. Hydrometallurgical processing involves solvents such as nitric acid, hydrochloric acid, cyanides and thiourea and these can lead to concomitant waste-management and environmental problems. The previously mentioned novel technologies may assuage these problems. Recovery rates of better than 84% for copper and better than 89% for lead have been obtained by using super-critical water oxidation and electrokinetic separation.

Given that the main interest in the recycling of e-waste is driven by the recovery of rare earths, the processes currently used in their recovery and separation are bioleaching, hydrometallurgy, pyrometallurgy, electrochemistry, siderophore technology and the use of carbon-based materials.

Recycling methods

Bioleaching and biosorption

Bioleaching is based upon the use of micro-organisms. It is conventionally used[43] in the recovery of antimony, arsenic, cobalt, copper, gallium, molybdenum, nickel, palladium, platinum, osmium and zinc from ore (table 8). Bioleaching is now applied to the extraction of metals from electronic waste, fly-ash and spent catalysts. As compared to hydrometallurgy and pyrometallurgy, bio-recovery is more cost-effective and is environmentally friendly; copper recovery from biomass, for example, produces fewer contaminants when compared with conventional metal-processing techniques.

Materials Research Forum LLC
https://doi.org/10.21741/9781644901793

Table 8. Recovery of metals by bioleaching using micro-organisms

Micro-Organism	Metal	Recovery Efficiency (%)
Acidophilic consortium	aluminium	88
Penicillium simplicissimum	aluminium	95
Thermoplasma acidophilum	aluminium	64
Sb. Thermosulfidooxidans	aluminium	91
Sb. Thermosulfidooxidans	aluminium	94
Sulfobacillus thermosulfidooxidans	copper	89
Gallionella sp.	copper	95
At. Thiooxidans	copper	94
Acidophilic consortium	copper	97
Leptospirillum ferrooxidans	copper	95
Acidithiobacillus thiooxidans	copper	98
Ferroplasma acidiphilum	copper	99
Penicillium simplicissimum	copper	65
P. chlororaphis	copper	52
Thermoplasma acidophilum	copper	86
Sb. Thermosulfidooxidans	copper	95
At. Ferrooxidans	copper	95
Pseudomonas plecoglossicida	gold	69
P. chlororaphis	gold	8
Sulfobacillus thermosulfidooxidans	nickel	81
At. Thiooxidans	nickel	89
Penicillium simplicissimum	nickel	95
Thermoplasma acidophilum	nickel	74
P. chlororaphis	silver	12
Sulfobacillus thermosulfidooxidans	zinc	83
At. Thiooxidans	zinc	90
Acidophilic consortium	zinc	92
Penicillium simplicissimum	zinc	95
Thermoplasma acidophilum	zinc	80
Sb. Thermosulfidooxidans	zinc	96

The microbial recovery of rare earths is achieved by exploiting microbial metabolic processes when dissolved aqueous solutions, from whence recovery involves bio-absorption or bio-accumulation by microbial cells. Mobilization of rare earths from the solid phase can include the biochemical processes of redoxolysis, acidolysis and complexolysis. The first of these is a two-step process:

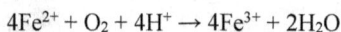

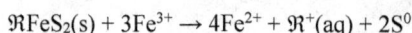

$$4Fe^{2+} + O_2 + 4H^+ \rightarrow 4Fe^{3+} + 2H_2O$$

$$\Re FeS_2(s) + 3Fe^{3+} \rightarrow 4Fe^{2+} + \Re^+(aq) + 2S^0$$

where $\Re$ is the rare earth and direct electron transfer from the mineral to the microbes occurs via the oxidation of Fe^{2+} to Fe^{3+}. Formation of the latter then results in oxidative dissolution of solid-phase rare earth. The micro-organisms which are involved in this process can include *Acidithiobacillus ferrooxidans, Acidithiobacillus thiooxidans and L. ferrooxidans*. In non-contact interactions, micro-organisms which are not attached to the mineral surfaces oxidize the dissolved Fe^{2+} to Fe^{3+}. This results, for example, in the oxidization of sulphide-type minerals to sulfuric acid and the dissolution of any rare earth. In contact interactions, the micro-organisms are in contact with a given mineral and generate so-called extracellular polymeric substances which surround microbial cells. The former are composed of polysaccharides, nucleic acids and proteins and are the reaction sites for the microbial oxidation of Fe^{2+} to Fe^{3+}, again followed by the oxidation of sulfides to sulfuric acid and the dissolution of any rare earths. The micro-organisms secrete organic acids in the extracellular polymeric substance layer and cause a reduction in pH and the dissolution of rare earths on mineral surfaces (table 9).

A strain of *A. thiooxidans* has been known to mobilize more than 99% of cerium, europium and neodymium, and 80% of lanthanum, from waste electronic equipment while *L. ferrooxidans, A. thiooxidans* and *A. ferrooxidans* are able to leach 100% of the praseodymium from powdered magnets. The bioleaching of neodymium by these three microbes is of the order of 91.3, 77.4 and 86.4%, respectively.

Table 9. Typical micro-organisms and the percentage leaching of rare earths

Source	Organism	T(C)	pH	Mechanism	Target	(%)
monazite	*Acidithiobacillus ferrooxidans*	30	1.8	redoxolysis	Ce	2
monazite	*Acidithiobacillus thiooxidans*	30	1.8	redoxolysis	Ce	2
monazite	*Acidithiobacillus ferrooxidans*	30	1.8	redoxolysis	La	1
monazite	*Acidithiobacillus thiooxidans*	30	1.8	redoxolysis	La	1
monazite	*Acidithiobacillus ferrooxidans*	30	1.8	acidolysis	Ce	9
monazite	*Acidithiobacillus thiooxidans*	30	1.8	acidolysis	Ce	9
monazite	*Acidithiobacillus ferrooxidans*	30	1.8	acidolysis	La	5
monazite	*Acidithiobacillus thiooxidans*	30	1.8	acidolysis	La	5
magnets	*Acidithiobacillus thiooxidans*	25	1.8	acido/redoxolysis	Pr	100
magnets	*Acidithiobacillus ferrooxidans*	25	3.2	acido/redoxolysis	Nd	86.4
magnets	*Leptospirillum ferrooxidans*	25	3.2	acido/redoxolysis	Pr	100
fly-ash	*Candida bombicola*	28	3.4	complexo/acidolysis	Yb	67.7
fly-ash	*Candida bombicola*	28	3.4	complexo/acidolysis	Er	64.6
fly-ash	*Candida bombicola*	28	3.4	complexo/acidolysis	Sc	63
ash slag	*Acidithiobacillus ferrooxidans*	45	4	acidolysis	Sc	52
ash slag	*Acidithiobacillus thiooxidans*	45	4	acidolysis	Sc	52
ash slag	*Acidithiobacillus caldus*	45	4	acidolysis	Y	52.6
ash slag	*Sulfobacillus sp.*	45	4	acidolysis	La	59.5

The reactions which are involved in acidolysis are such as to allow acid production to be mediated by sulfur-oxidizing and phosphate-solubilizing bacteria:

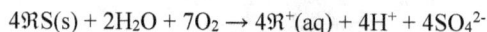

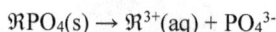

$$4\Re S(s) + 2H_2O + 7O_2 \rightarrow 4\Re^{+}(aq) + 4H^{+} + 4SO_4^{2-}$$

$$\Re PO_4(s) \rightarrow \Re^{3+}(aq) + PO_4^{3-}$$

The former bacteria oxidize sulfides to sulfuric acid, resulting in the dissolution of rare earths. Those sulfur-oxidizing micro-organisms which are capable of such acidolysis-mediated dissolution of rare earths include *A. ferrooxidans*, *A. thiooxidans*, *Alicyclosbacillus disulfidooxidans* and *S. acidophilus*.

The second reaction above involves phosphate-solubilizing micro-organisms which liberate phosphates from minerals and result in the solubilization of rare earths. The phosphate dissolution can occur via acidification by H^+ or organic-acid secretion, with the latter route leading to three-fold higher phosphate solubilization than does the former. Solubilizing micro-organisms which are able to dissolve mineral phosphate include *Acetobacter, Acidithiobacillus, Enterobacter, Erwinia, Flavobacterium, Klebsiella, Micrococcus, Pseudomonas, Rhizobium, Serratia* and *Streptomyces*. They are able to secrete organic acids such as acetic, citric, gluconic, 2-ketogluconic, malic, oxalic and succinic. The type of secretion depends upon the type of organic carbon used for microbe growth, thus *Acinetobacter sp.* produces gluconic acid when glucose is the carbon source but produces malic acid when mannitol is the source. The ability to produce organic acids and to solubilize phosphate minerals also depends upon the composition of the growth medium, with *Pseudomonas aeruginosa* producing only gluconate under phosphate-deficient conditions but producing pyruvate and citrate as well as gluconate under phosphate-rich conditions. Enzymes are also released which contribute to the solubilization of rare earths, with phosphatase from *Enterobacter aerogenes*, *Pantoea agglomerans* and *Pseudomonas putida* being able to solubilize phosphate and thus bioleach rare earths from monazite rock.

Complexolysis involves the production of microbial metabolites such as organic acids and siderophores. Acetic, citric, fumaric, lactic, malic, oxalic and succinic acids, produced by *Acetobactor sp.,* can extract more than 50% of yttrium and scandium from some sources via the formation of complexes of rare earths with organic acids. Media which contain gluconic acid produced by *Gluconobacter oxydans* can extract yttrium, samarium, ytterbium, europium, neodymium and cerium, to the extent of 91.2, 73.2, 83.7,

50, 42.8 and 36.7%, respectively, from synthetic phosphogypsum at a pH level of 2.1. This is attributed to the formation of a complex of gluconate with the rare earth.

The bio-accumulation ability of micro-organisms is affected by acidic conditions because of a poor interaction between rare earths and the proton-saturated functional groups on the surface of the cell. The accumulated rare earths in fact act as metabolic factors in the microbe metabolism. It seems for example that *Methylacidiphilum fumariolicum* can use europium, gadolinium, lanthanum, neodymium, praseodymium and samarium as co-factors for methanol dehydrogenase during methane oxidation. Many strains belonging to the orders, *Burkholderiales, Caulobacteriales, Methylococcales, Neisseriales, Rhizobiales, Rhodobacteriales, Rhodospirillales, Vibrionales* and *Xanthomonadales,* indeed possess methanol dehydrogenase enzymes and thus may depend upon rare earths for their metabolic requirements.

The recovery of dissolved rare earths is possible via micro-organism controlled bioprecipitation in which the release of inorganic phosphates during microbial metabolization leads to the precipitation of rare earths in the form of phosphates:

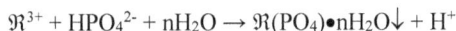

$$\mathfrak{R}^{3+} + HPO_4^{2-} + nH_2O \rightarrow \mathfrak{R}(PO_4)\bullet nH_2O\downarrow + H^+$$

$$\mathfrak{R}^{3+} + H_3PO_4 + nH_2O \rightarrow \mathfrak{R}(PO_4)\bullet nH_2O\downarrow + 3H^+$$

For example, the phosphatase enzyme within *Serratia sp.* generates inorganic phosphates and precipitates dissolved rare earths from aqueous solution, so that over 90% of neodymium and 85% of europium can be recovered by phosphate-based precipitation using *Serratia sp.* on polyurethane foam.

The bioprecipitation process is affected by the pH value because of the solubility of the rare earth phosphate precipitate. The recovery of lanthanum, in the form of $LaPO_4$, by *Citrobacter sp.* phosphorylation is reduced to 50%, by a pH level of 5, because of an insufficient de-solubilization of $LaPO_4$ under acidic conditions. The nature of the bioprecipitation process permits the recovery of specific rare earths. Thus the treatment of fluorapatite with *B. megaterium* involves the selective dissolution of heavy rare earths and the precipitation of light rare earths in the form of phosphate salts.

Oxygen generally acts as the terminal electron-acceptor for the metabolism of acidophilic chemolithotrophic micro-organisms which mediate rare-earth biorecovery, and thus aeration is an important factor in the growth of the associated micro-organisms. Carbon dioxide and oxygen are the carbon source and electron acceptor, respectively, for the autotrophic micro-organisms *A. ferrooxidans* and *A. thiooxidans*, and therefore should be plentifully supplied by aeration during bioleaching since this improves the bioleaching efficiency of *A. ferrooxidans*. The pulp density is also an important factor because a high density affects microbial growth by increasing shear forces and limiting the ingress of carbon dioxide and oxygen. The pulp density also affects the pH value, due to its high buffering capability. The optimum pulp density for maximizing rare earth biorecovery depends upon the source and upon the type of microbe. An optimum pulp density of 2%w/v, when using *Acetobacter sp.*, leads to a 61% recovery of yttrium and to a 52% recovery of scandium. A higher pulp density (10%w/v) is required for the optimum recovery of 52% of scandium, 52.6% of yttrium and 59.5% of lanthanum from coal ash when using a mixture of *A. ferrooxidans*, *Leptospirillum ferriphilum* and *S. thermosulfidooxidans*.

Temperature plays an important role in the biorecovery because of its direct effects upon microbial growth and metabolic activity. On the basis of their temperature requirements, micro-organisms can be psychrophilic, mesophilic or thermophilic, with optimum growth temperatures of -4 to 20C, 25 to 47C and 41 to 68C, respectively (table 10).

The oxidative capability of microbial groups depends upon that optimum growth temperature, with *A. ferrooxidans* having an optimum growth temperature of 28 to 30C and *Acidithiobacillus ferrivorans* and *Acidithiobacillus caldus* having optimum growth temperature of 5 and 45C, respectively. Use of the optimum growth temperature also increases the attachment of microbial cells to mineral surfaces, with the attachment of *A. ferrooxidans* to the surface of pyrites increasing four-fold when the temperature is increased from 17 to 28C. *L. ferrooxidans* can recover 100% of the dysprosium and praseodymium from scrap magnets when grown at 25C.

The growth of the micro-organisms involved in rare-earth biorecovery is meanwhile largely dependent upon the pH level of the aqueous solution, with the micro-organisms which are involved in the biorecovery generally being acidophilic. The fungal strain of *Candida bombicola* extracts 67.7% of the ytterbium, 64.6% of the erbium and 63% of the scandium from coal fly-ash at a pH level of 3.3 to 3.5. A highly acidic pH value leads to a stronger attachment of micro-organisms to mineral surfaces and to improved rare earth recovery. Adsorption-based recovery is also affected by the pH level, with the rare-earth adsorbing groups on microbial-cell surfaces being occupied mainly by H^+ under acidic conditions and resulting in lower rare-earth adsorption.

Table 10. Optimum conditions for the micro-organism bioleaching of rare earths

Micro-organism	Optimum Temperature (C)	Optimum pH
Acidithiobacillus ferriredurans	29	2.1
Acidithiobacillus ferrivorans	27–32	2.5
Acidithrix ferrooxidans	25	3.0-3.2
Ferrimicrobium acidiphilum	35	2
Ferrithrix thermotolerans	43	1.8
Aciditerrimonas ferrireducens	50	3
Alicyclobacillus aeris	30	3.5
Alicyclobacillus ferrooxydans	28	3
Ferrovum myxofaciens	32	3
Thiomonas islandica	45	4
Sulfobacillus thermotolerans	40	2
Sulfobacillus benefaciens	38–39	1.5
Ferroplasma cupricumulans	53.6	1–1.2
Ferroplasma thermophilum	45	1
Acidianus manzaensis	74	0.8–1.4
Metallosphaera cuprina	65	3.5

The dissolution of rare earths via micro-organism mediated redoxolysis depends upon the redox potential of the aqueous environment. The oxidation of Fe^{2+} to Fe^{3+} by bacteria such as *A. ferrooxidans*, *A. thiooxidans* or *L. ferriphilum* occurs at a very high redox potential with, in the case of the above oxidation by *A. ferrooxidans*, the potential apparently being greater than 600mV during the bioleaching of waste light-emitting diode lamps. Extraction of 52% of the scandium, 52.6% of the yttrium and 59.5% of the lanthanum from coal slag, using *A. caldus*, involves a redox potential of 845 to 855mV. Sulfur oxidation by *Sulfobacillus sp.* is expected to occur efficiently at a redox potential of 100 to 150mV.

The recycling of scrap magnets having various alloy compositions and particle sizes has been achieved[44] by bioleaching using bacteria. The highest leaching efficiencies were obtained by using *Acidithiobacillus* and *Leptospirillum ferrooxidans*, with leaching efficiencies of up to 100% in the case of dysprosium and praseodymium. The leaching efficiencies of non-bacterial control treatments were of the same order, and chemical leaching by acids could be superior. On the other hand, batches which included iron additions led to better leaching due to the catalytic effect of the Fe^{3+} ions. Bioleaching was more efficient overall due to a lower cost, fewer chemicals and lower pollution arising from emissions and residues. No loss in efficiency was noted upon scaling-up the process. Further purification was best done by precipitation with oxalic acid in a two-step process, and rare-earth extraction rates of up to 100%, with a purity of 98%, was possible. Bioleaching has also been applied[45] to the recycling of fluorescent phosphors by using a wide range micro-organisms, including both acidophilic and heterotrophic ones. Larger amounts of rare earths were found to be leached by *Komagataeibacter xylinus*, *Lactobacillus casei* and *Yarrowia lipolytica*. A COOH-functionality and other biotic processes contributed to leaching. Among the various rare-earth components of the scrap was the red dye, yttrium europium oxide. A two-step bioleaching process has been developed[46] for the recovery of various metals, including rare earths, from the dust which is generated by the shedding of waste electrical and electronic equipment. In the first step, base metals are almost totally leached from the dust by using *Acidithiobacillus thiooxidans* over a period of 8 days. This lowers the pH of the leaching solution from 3.5 to 1.0 and, during this step, cerium, europium and neodymium are extracted to better than 99%, while lanthanum and yttrium are extracted to the extent of 80%. In the second step, cyanide-producing *Pseudomonas putida* extracts 48% of gold - within 3h - from the *Acidithiobacillus thiooxidans* leachant. Because the separation of the various components of a commercial product is a highly labour-intensive operation, the leaching process has been extended into the concept of so-called biodismantling[47]. There had previously been no systematic method for dismantling and recycling electronic components. The new technique uses bioleaching to reduce costs and environmental harm. Scrap printed circuit boards have been subjected to bioleaching by using an iron-oxidizing culture comprising mainly *Acidithiobacillus ferrooxidans*, plus 20mM of ferrous iron sulphate, with a pH of 1.8 for 20 days at 30C. Non-bacterial control sample were treated under similar conditions, with either 20mM of Fe^{2+} or 15mM of Fe^{3+}. After 20 days of treatment, dismantling was achieved by using either the bioleaching or the Fe^{3+} control mixture. The Fe^{2+} control mixture was ineffective. The bioleaching mixture led to a lower rate of dismantling, but this was attributed to a constantly higher redox potential and a consequent competition with solder leaching and copper leaching from the printed copper wires.

Biosorbents are environmentally friendly, cheap and plentiful but require a very high adsorption capacity. *Pseudomonas sp.*, with an absorption capacity of 950mg/g and *Monoraphidium sp.* (1511mg/g) are among those recommended for the pre-concentration of rare earths. Tetra-octyldiglycolamide is perhaps the most promising ligand for the liquid-liquid extraction of rare earths. Adsorbents such as tetra-octyldiglycolamide-modified carbon inverse opals and tetra-octyldiglycolamide-functionalized carbon nanotubes have very high lutetium/lanthanum separation factors of 794 and 690, respectively, but the adsorption capacity of the former is only 18mg/g. Biosorbents have higher adsorption capacities, but high separation factors for rare earths are not assured.

Table 11. Recovery of rare earths by biosorption

Biosorbent Material	Metal	Efficiency (mg/g)
Bacillus cereus	silver	91.75
Bayberry tannin	palladium	33.4
Chlorella vulgaris	lanthanum	74.60
Chlorella vulgaris	praseodymium	157.21
Klebsiella sp.	silver	141.1
Platanus orientalis	cerium	32.05
Pleurotus ostreatus basidiocarps	yttrium	54.54
Racomitrium lanuginosum	gold	37.2
Sargassum sp.	lanthanum	91.68
Sargassum sp.	praseodymium	98.63
Turbinaria conoides	cerium	146.4
Ulva lactuca	cerium	69.75

Extensive studies have been made of the recovery of rare earths from e-waste by using algae, bacteria and fungi as biosorption materials (table 11). Lanthanum has been recovered by using *Sargassum* biomass, *Pseudomonas sp.* and *Agrobacterium sp.* as an adsorbent. In the case of neodymium, *Monoraphidium sp.*, bakers' yeast, *Penicillium sp.*, *Saccharomyces cerevisiae*, *Kluyveromyces marxiamus*, *Candida colliculosa* and *Debaryomyces hansenii* have been used. Cerium can be biosorbed by *Platanus orientalis* leaf powder and *Agrobacterium sp*. These methods are very efficient and cost-effective in recovering metals from solution. The biosorption of metals depends upon the temperature of 25 to 50C, the pH-value of 4 to 7.5 and the agitation rate at initial metal concentrations of 15 to 300mg/l. Biosorption is a promising low-cost technique for the recovery of rare earths from electronic waste. Rare earths have been recovered[48] from waste electrical and electronic equipment by using a strain of *penicillium expansum* which was isolated from an ecotoxic metal-contaminated site. The resultant product was a highly concentrated solution of lanthanum (up to 390ppm) and terbium (up to 1520ppm).

The distribution of active sites on the biosorbent's surface governs the overall adsorption ability. A very large surface area thus favours the adsorption of metal ions. At the same time, the pore distribution controls the internal diffusion of metal ions and their accumulation within the biosorbent. The Brunauer–Emmett–Teller surface area in m^2/g, and the pore volume in cm^3/g, are thus critically important parameters (table 12).

The BET surface area is low for biosorbents other than those which contain silica or carbon, thus suggesting that the particle size is smaller in those biosorbents. The pre-treatment of raw *Saussurea tridactyla* using NaOH leads to a 30-fold increase in the BET surface area, but to only a doubling of the adsorption capacity. It is difficult to discern a definite correlation between the surface area and the adsorption capacity of a biosorbent. The adsorption capacity is affected by the adsorbent particle-size, with an increase in particle size – and a concomitant decrease in the surface area – lowering, for example, the adsorption capacity of bamboo-derived carbon.

Table 12. Properties of selected biosorbents

Biosorbent	BET Surface Area (m^2/g)	Pore Volume (cm^3/g)
chitosan–silica	218.0	1.024
prawn carapace	56.3	0.170
corn style	47.3	0.120
fish scales	23.3	0.120
neem sawdust	13.7	0.070
ion-imprinted membrane material	3.2	0.011
raw *Saussurea tridactyla*	1.1	0.005
chitosan	1.0	0.003

The adsorption of a rare earth by a biosorbent depends upon the pore size and volume. The entry of rare-earth ions into pores, followed by their diffusion is more feasible when the pore size is larger than the size of the hydrated ion. The greater the pore volume, the greater the number of rare-earth ions that can be accommodated; thus leading to a higher adsorption capacity.

The pH of a solution plays an important role in the adsorption process because it strongly affects the speciation of metal ions in the solution and the surface polarity of the biosorbent. Many biosorbents possess one or more carboxylic, amine, hydroxyl or thiol groups and their acid-base nature depends upon the pH value and modifies the affinity of those functional groups for other metal ions. The biosorption of rare earths on various biosorbents tends to be studied for pH values ranging from 1 to 7 because, at higher pH values, hydrolysis of the rare earth ions leads to the formation of insoluble hydroxides. Different chemical species of rare earth ions predominate in aqueous solution at differing pH values. There can be an increase in the adsorption of rare earths by biosorbents with increasing solution pH, and this is attributed to the de-protonation of functional groups on the biosorbent surface, with increasing pH, resulting in the availability of binding sites for metal-ion adsorption. At lower pH values, protonation of those functional groups is favourable and the adsorption of positively-charged metal ions is then hindered due to electrostatic repulsion between the ions and the protonated functionalities. With increasing pH, the latter begin to de-protonate; resulting in the creation of negatively-charged functionalities which possess a greater affinity for positively charged are-earth ions.

There always exists an optimum concentration of the adsorbent which can lead to the best adsorption performance by maximizing the interaction between metal ions and the binding sites of the biosorbent. An increasing concentration of adsorbent favours good adsorption performance and is equivalent to an increase in the number of active sites. In general, the adsorption performance saturates at some biosorbent dosage level before dropping off rapidly. When Pr^{III} is adsorbed on *T. arjuna* bark powder, the adsorption has been known to improve from 42 to 87% when the biosorbent dose is increased from 0.01 to 0.03g/l.

In a very few cases, the percentage adsorption efficiency decreases above a certain biosorbent dosage. An adsorbent dose above that threshold can render active sites inaccessible to the metal ions because of so-called sorbate-sorbate interactions. This decreases the adsorption capacity, and this anomalous behaviour has been termed the cesium-effect. It is observed in the adsorption of trivalent rare earths on functionalized multi-walled carbon nanotubes. The adsorbent dose is in fact a distribution which arises from variations in the functionalities of biosorbents having various origins.

Materials Research Forum LLC
https://doi.org/10.21741/9781644901793

The ionic-strength dependence yields information on complexation mechanisms and two different mechanisms, known as inner-sphere and outer-sphere, are involved in the adsorption of metal ions onto the adsorbent surface. If the adsorption capacity decreases with increasing ionic strength, the process involves an outer-sphere electrostatic complexation mechanism. In the case of the inner-sphere complexation mechanism (ligand exchange), the adsorption performance may increase or remain unchanged with increasing electrolyte concentration. In the case of the biosorption of rare earths onto phosphorylated cactus fibres, pre-treated crab shell particles or magnetic alginate-chitosan gel beads, an increase in ionic strength weakens adsorption, thus showing that outer-sphere complexation governs the adsorption process. Increasing the ionic strength introduces a shielding effect for the interaction of rare-earth ions with functionalities on the biosorbent surface.

The surface of peanut shells has been grafted with polyacrylic acid by using ultra-violet technology and the modified shells were tested for their Ce^{3+} adsorption capability, showing that the ideal adsorption capacity occurred at pH levels of 4 to 7 and was slightly affected by ionic strength[49]. The Langmuir model was better than the Freundlich model for describing the adsorption of Ce^{III}; which obeyed pseudo second-order kinetics. The maximum adsorption capacity was estimated to be 134.59mg/g, and the absorption capacity remained above 88% during 5 consecutive cycles. A syringe-like adsorption device[50] has used diglycolamic-acid modified chitosan sponges as adsorbents for the recycling of rare earths. By combining the elasticity of sponges with the selective extraction behaviour of diglycolamic-acid groups, the device could efficiently extract rare earths from aqueous solution. It required just 600s to achieve adsorption equilibrium, and squeezing the sponges produced solid-liquid separation. This method avoided the pollution caused by organic solvents, shortened the time required for adsorption equilibrium and could recover yttrium and europium from waste phosphors.

Various biosorbents have been considered for the separation of rare earths, but few have exhibited the required rare-earth adsorption capacity (table 13). Comparison of the biosorption performance of unmodified and modified biosorbents shows that the latter are much more efficient in adsorbing rare earths.

Materials Research Forum LLC
https://doi.org/10.21741/9781644901793

Table 13. Modified biosorbents for the adsorption of trivalent rare earths

Biosorbent	Rare Earth	Adsorption Capacity (mg/g)
silica-chitosan hybrid[a]	yttrium	158.8
Ankistrodesmus sp.	lanthanum	1250.1
carboxymethyl cellulose[b]	cerium	320.5
kenaf cellulose-based ligand[c]	praseodymium	244.0
Sacran-sepiolite bio-nanocomposite film	neodymium	1932.8
kenaf cellulose-based ligand[d]	samarium	192.0
magnetic chitosan micro-particles[e]	europium	375.4
chitosan/magnetite nanocomposite[f]	gadolinium	1483.6
NaOH treated *Saussurea tridactyla*	terbium	149.0
Penidiella sp. T9	dysprosium	910.0
NaOH treated *Saussurea tridactyla*	holmium	117.0
glutaraldehyde crosslinked chitosan	erbium	124.0
NaOH treated *Saussurea tridactyla*	thulium	153.0
NaOH treated *Saussurea tridactyla*	ytterbium	138.0
NaOH treated *Saussurea tridactyla*	lutetium	114.0

a: 3-aminopropyl triethoxysilane/1–(2-pyridylazo) 2-naphthol modified silica-chitosan hybrid, b: monolithic open-cellular hydrogel adsorbents based on carboxymethyl cellulose, c: kenaf cellulose-based poly(amidoxime) ligand, d: amidoxime-functionalized magnetic chitosan micro-particles, e: diethylenetriaminepentaacetic acid-functionalized chitosan/magnetite nanocomposite, f: magnetic poly(aminocarboxymethylation) functionalized glutaraldehyde cross-linked chitosan

The effect of other metal ions upon the selective adsorption of rare earths using biosorbents has been studied in some detail. The separation of rare earths from leachates having a low (nmol/l) concentration, using so-called lanthanide binding-tagged *E. coli* strains reveals the preferential adsorption of rare earths over that of several non rare earth metals. The lanthanum binding which was introduced into the *E. coli* increased the adsorption capacity by between 2 and 10 times and made it select rare earths over non rare earths and enhanced the affinity for rare earths as a function of decreasing atomic radius. An exception is copper, which is revealed to be a strong competitor to the rare earths. The biosorption of lanthanum, europium and ytterbium on *P. aeruginosa* also markedly decreases in the presence of Al^{3+} ions as compared with the effect of monovalent (potassium, sodium) or divalent (calcium) ions. This is attributed to the similarity of the ionic charges. A higher uptake of trivalent iron or neodymium, among other trivalent metal ions, has been tentatively attributed to their lower dehydration energies. This would ease a partial stripping of the water molecules from their hydration spheres before their adsorption on an alginate, for example. Trivalent metal ions also interact strongly with the COO– groups in alginate, as compared with divalent transition-metal ions, and are therefore more highly adsorbed. An *Escherichia coli* strain has been encapsulated[51] in a permeable polyethylene glycol diacrylate hydrogel, at high cell-density, by using an emulsion process. This resultant bead-like adsorbent contained an homogenous distribution of cells with surface functional groups. The beads were packed into fixed-bed columns, and could provide effective neodymium extraction at a flow rate of up to 3m/h in pH levels of 4 to 6. The columns were stable for re-use and retained 85% of their adsorption capacity after 9 consecutive adsorption/desorption cycles.

Polyethylenimine-acrylamide/SiO_2 hybrid hydrogels have been used[52] as sorbents, for rare-earth recycling from aqueous sources, because of their remarkable stability and selectivity. The organic-inorganic hybrids were synthesized via the thermopolymerization of acrylamide onto polyethylenimine polymer chains, with N,N′-methylene bis(acrylamide) as a cross-linker. The pH level of the medium greatly affects rare-earth elements adsorption, with almost neutral conditions ensuring the strongest bonding of rare earths to active sites. The rate-determining step of the adsorption process is chemical sorption, and rare-earth diffusion within micropores is the control step for intraparticle diffusion. A high selectivity of rare earths over interfering metals was possible by using a citrate-based buffer eluent.

Due to the similar properties of the rare earths, their separation by using biosorbents is difficult. The so-called rare-earth selectivity ratio can help to predict the adsorption preferences of a given biosorbent. Recalling that the surface of a biosorbent exhibits many different non-uniformly distributed functionalities, a high degree of discrimination between rare earths is not to be expected. In most cases, the selectivity ratio ranges from 1 to 2 (table 14). On the other hand, a selectivity ratio of 25 has been observed for rare-earth adsorption on alginic acid gel-beads, and a ratio of 13.7 found for cerium over lanthanum in the case of bayberry tannin grafted chitosan, thus opening up the possibility of separating the light rare earths from the heavy ones. A strain of *P. expansum* can moreover concentrate lanthanum and terbium to between 200 and 3000 times their initial concentrations even when silver, aluminium, gold and iron are present in concentrations of 5000 to 72000ppm.

Table 14. Selectivity ratios of bio-absorbents for rare earths

Rare-Earth Pair	Bio-Absorbent	Selectivity Ratio
ytterbium/dysprosium	diethylenetriamine-functionalized chitosan	1.00
lanthanum/cerium	*Turbinaria Conoides*	1.01
lanthanum/cerium	brown marine algae	1.02
thulium/terbium	*Saussurea tridactyla*	1.03
yttrium/cerium	intercalated cellulose nanocomposite	1.06
ytterbium/neodymium	diethylenetriamine-functionalized chitosan	1.07
lanthanum/cerium	grapefruit peel	1.08
yttrium/lanthanum	intercalated cellulose nanocomposite	1.10
gadolinium/lanthanum	banana peel	1.10
ytterbium/neodymium	cysteine-functionalized chitosan	1.11
thulium/ytterbium	*Saussurea tridactyla*	1.11
lanthanum/europium	*Turbinaria Conoides*	1.12
cerium/lanthanum	*Platanus orientalis*	1.12
thulium/dysprosium	*Saussurea tridactyla*	1.13
lanthanum/samarium	orange waste gel	1.14
neodymium/lutetium	diglycolic amic acidmodified *E. coli*	1.15
neodymium/dysprosium	diglycolic amic acidmodified *E. coli*	1.16
thulium/lanthanum	*Saussurea tridactyla*	1.18
thulium/gadolinium	*Saussurea tridactyla*	1.18
lanthanum/holmium	orange waste gel	1.18
holmium/erbium	alfalfa biomass	1.19
lanthanum/neodymium	*parachlorella*	1.20
lanthanum/neodymium	soy hull	1.21
lanthanum/europium	cellulose based silica nanocomposite	1.21

thulium/europium	*Saussurea tridactyla*	1.21
lanthanum/europium	brown marine algae	1.22
lanthanum/ytterbium	*Pseudomonas aeruginosa*	1.22
ytterbium/lanthanum	cysteine-functionalized chitosan	1.26
yttrium/europium	*Gracilaria gracilis*	1.28
lanthanum/ytterbium	*Turbinaria Conoides*	1.28
yttrium/cerium	*Gracilaria gracilis*	1.30
thulium/holmium	*Saussurea tridactyla*	1.31
lanthanum/cerium	*Pinus brutia*	1.33
thulium/erbium	*Saussurea tridactyla*	1.33
thulium/lutetium	*Saussurea tridactyla*	1.34
thulium/neodymium	*Saussurea tridactyla*	1.35
lanthanum/europium	*Pseudomonas aeruginosa*	1.37
yttrium/neodymium	*Gracilaria gracilis*	1.41
thulium/praseodymium	*Saussurea tridactyla*	1.47
thulium/samarium	*Saussurea tridactyla*	1.48
thulium/cerium	*Saussurea tridactyla*	1.51
lanthanum/ytterbium	brown marine algae	1.59
yttrium/lanthanum	*Gracilaria gracilis*	1.79
lanthanum/samarium	soy hull	1.83
gadolinium/neodymium	alfalfa biomass	1.98
gadolinium/cerium	chitosan/carbon nanotube composite	2.23
lanthanum/samarium	alginic acid gel beads	2.30
thulium/yttrium	*Saussurea tridactyla*	2.46
cerium/europium	crab shell	2.93
lanthanum/praseodymium	*parachlorella*	3.47
gadolinium/lanthanum	chitosan/carbon nanotube composite	3.50

lanthanum/yttrium	*parachlorella*	5.13
lanthanum/samarium	*parachlorella*	6.55
lanthanum/gadolinium	*parachlorella*	11.35
cerium/lanthanum	bayberry tannin grafted chitosan	13.70
lanthanum/dysprosium	*parachlorella*	16.12
lanthanum/ytterbium	alginic acid gel beads	25.00

Siderophores

Siderophores are also related to micro-organisms in that they are small iron-chelating compounds which are secreted by bacteria, fungi and grasses. The chelating molecules have apparently evolved in order to harvest Fe^{3+} ions from the environment, for use by the micro-organism and are consequently the best ligands for ferric ions. The siderophores can also exhibit a high affinity for other metals, such as rare earths, and for desferrioxamine. A higher recovery efficiency has been observed for lithium and molybdenum, and a lower efficiency for cerium; due mainly to the formation of cerium complexes with siderophores which harboured competing cations such as Fe^{3+}. The recovery of rare earths by using siderophores is cost-effective, rapid and environmentally friendly when compared with other methods.

Desferrioxamine is a well-known siderophore which serves as an organic chelating agent for the extraction of rare earths from minerals. Bacteria belonging to the *Actinobacteria* metal-binding strain are often present in rare-earth bearing rocks and can produce siderophores. Siderophores which are provided by *Aspergillus niger* permit the bioleaching of 51% of the lanthanum and 50.1% of the cerium from phosphorite at a pH level of 7 and a temperature of 30C. The production of siderophores can be increased by substrates such as glucose and glycerol, and the use of such a substrate may greatly improve the yield of rare earth extraction during the bioleaching process.

Adsorption is controlled by cell-surface functional groups such as -COOH, -NH$_2$ and HPO$_4^{2-}$. In the case of *Thermus scotoductus,* PO_4^{3-}, $C-PO_3^{2-}$, COOH and C=O groups on its surface are implicated in an electrostatic interaction which promotes the specific adsorption of europium. The cell-surface functional group, $-PO_4^{3-}$, exhibits a specificity for gadolinium and -COOH exhibits a specificity for ytterbium, erbium and samarium. Lanthanum and neodymium have an affinity for $-PO_4^{3-}$ while samarium, gadolinium, erbium and ytterbium exhibit a similar specificity with respect to both the -COOH and -PO_4^{3-} functional groups. The binding of rare earths to $-PO_4^{3-}$ groups occurs at pH levels of 3 to 4, while binding to -COOH groups occurs at pH levels of 6 to 7. The adsorption of rare earths depends upon their molecular weight, with light members such as lanthanum and neodymium being adsorbed efficiently at pH levels greater than 4, and other members such as samarium and gadolinium being strongly adsorbed at pH levels below 4.

Micro-organisms such as the fungus, *Saccharomyces cerevisiae*, the Gram-positive bacterium, *Bacillus subtilis*, and the Gram-negative bacterium, *Pseudomonas fluorescens*, exhibit similar adsorption mechanisms for samarium. That is, adsorption begins with an interaction between the samarium and cell-surface organic molecules, including phospholipids, polyphosphates and polysaccharides, to form nucleation sites when phosphorylation reactions generate samarium-bearing phosphate precipitates. Cerium adsorption similarly occurs via inner-surface complex formation between the cerium and functional groups, following the formation of cerium-phosphate nanoparticles on the cell surface. Light, medium and heavy (molecular weight) rare earths exhibit differing types of binding to functional groups on the microbial cell surface. The total binding-site concentration on the surface of *B. subtilis* decreases, for instance, in going from light earths such as lanthanum, cerium and neodymium to heavy earths such as thulium, ytterbium and lutetium. This is attributed to the multi-dentate binding of heavy rare earths as compared with that of light rare earths, which prefer mono-dentate binding.

The capacity and specificity of rare-earth biosorption can be increased by modifying the micro-organisms so that they acquire lanthanide binding tags on their surface. Modified *Caulobacter crescentus*, with eight copies of lanthanide binding tags on its surface, exhibits 50, 46.7, 36 and 35.5% increases in the biosorption - from ore leachate - of yttrium, lanthanum, cerium and neodymium, respectively, as compared with that of control samples bearing no lanthanide binding tags. In the same way, *Escherichia coli*, bearing 16 copies of lanthanide binding tags on its surface, exhibits increases of 56.1, 87, 72.4, 63.6 and 58.8% in the biosorption of yttrium, lanthanum, praseodymium, cerium and neodymium, respectively, as compared with that of control samples.

Bio-accumulation involves the intracellular up-take of rare earths which are adsorbed on the cell surface. Following adsorption at the cell surface, so-called importer complexes in the membrane lipid bi-layer transports the rare earth into intracellular space, where they are sequestered by proteins and peptide ligands. Many micro-organisms can bio-accumulate rare earths from the environment, as in the bio-accumulation of cerium and neodymium by *Bacillus cereus* in rare-earth enriched soil, where exposure to cerium increases the appearance of -COOH groups on the bacterial cell-surface.

A newly discovered protein, Lanmodulin, undergoes large conformational changes in the presence of rare earths and has been shown to be the most selective macromolecule with respect to rare earths. It exhibits exceptional rare-earth binding down to pH levels of about 2.5, and its rare-earth complexes are stable at up to 95C. It has been used[53] to extract rare earths pre-combustion coal and electronic waste leachates. Following a single all-aqueous step, the rare earths had been selectively recovered from among lithium, sodium, magnesium, calcium, strontium, aluminium, silicon, manganese, iron, cobalt, nickel, copper, zinc and uranium ions.

On the other hand, the low solubility of rare earths in lipids can generally limit their bio-accumulation, although *Arthrobacter luteolus* evades this limitation via the production of a catechol-type siderophore that forms a rare-earth siderophore complex and accumulates the complex through its membrane. A principal advantage of bio-accumulation is the possibility of the selective recovery of a given rare earth from a mixed-metal solution, and the attraction of rare earths to phosphoryl ligands permits, for example, the acid-tolerant micro-algae, *Galdieria sulphuraria,* to recover more than 90% of neodymium, dysprosium and lanthanum.

Bacteriophages have been used as biocollectors in a model bioflotation system for the separation of lanthanum phosphate doped with cerium and terbium ($LaPO_4:Ce^{3+},Tb^{3+}$) from mixed fluorescent phosphors[54]. A phage surface display was used to develop peptides having a high specificity for particular targets in electronic scrap. A phage clone containing a particular surface peptide loop was found to bind specifically to the test material. Binding and immunofluorescence assay confirmed a peptide affinity for $CeMgAl_{11}O_{19}:Tb^{3+}$ and $BaMgAl_{10}O_{17}:Eu^{2+}$, while there was no affinity for other fluorescent phosphor compounds such as $Y_2O_3:Eu^{3+}$. The binding specificity of the original peptide loop could be improved by over 50 times by using alanine scanning mutagenesis[55].

The adsorption ability of *spirulina* powder has been investigated[56] with regard to the recovery of ytterbium from waste-water. The surface structure and valence of the adsorbent were analyzed in order to determine the adsorption mechanism. The characteristics of Yb^{III} on spirulina powder were assessed by using adsorption isotherm, kinetic and thermodynamic models. The adsorption isotherm data could be explained by the Langmuir model, with the Yb^{III} adsorption capacity of the powder being 72.46mg/g at 318K. A pseudo second-order kinetic model simulated the Yb^{III} adsorption ability of *spirulina* powder, suggesting that the rate-controlling step is chemical adsorption.

Magnetic adsorbents have been created[57] by directly grafting functional ligands onto the surface of iron-oxide nanoparticles. Each of the ligands carried 2 functional groups, including phosphonic acid plus either alcohol, thiol, amino, carboxylic- or phosphonic acid. The characteristics of metal-cation adsorption were determined for Sm^{3+}, Nd^{3+}, Dy^{3+}, Tb^{3+}, Co^{2+} and Ni^{2+}; these being representative of the magnetic-material constituents which might require recycling. Long (C_{10} to C_{12}) linear alkyl chains were used to estimate potential stabilization of the particles against iron leaching during the desorption of cations in acidic media. In the case of phosphonic acid, a C_6 chain was sufficient. The nanoparticles exhibited a high adsorption capacity, efficient desorption and an appreciable selectivity between rare earths and late transition metals, while the ligands protected the magnetic particles against iron leaching.

Algae and seaweed

Algae-based methods are one of the most promising methods for recovering rare earths. This is because of its high efficiency, low cost and wide applicability. Microalgae of the *Coccomyxa* genus, typically isolated from extreme environments, can withstand radiation and other stresses and also exhibit non-selective metal up-take. This makes them choice organisms for the development of bioprocesses for rare-earth extraction. Unicellular red algae, *Galdieria phlegrea*, have been used[58] as an experimental organism with which to examine the bio-accumulation of rare earths[59] from luminophores. The cells were cultured mixotrophically in a liquid medium with added glycerol as a carbon source. Luminophores from energy-saving light bulbs and fluorescence lamps were then added to the medium in powdered form. The total rare earth content was twice as high in the scrap from bulbs as that from lamps. The most abundant element, circa 90% by weight, in both cases was yttrium. The growth of cultures in the presence of the luminophores was increased, especially in the case of lamp scrap. The most abundant element that accumulated in the algal biomass was yttrium, followed by europium and lanthanum. The chlorophyll content of the algae was markedly increased by the luminophores, and more so in the case of lamp scrap. Various strategies for improving the rare-earth removal efficiency by dried and live algae have been developed[60], depending upon the properties of the dried or live algae.

Live seaweed has been shown to be able to remove rare earths from contaminated solutions. This ability was analyzed[61] by exposing 6 such seaweeds (*ulva lactuca, ulva intestinalis, fucus spiralis, fucus vesiculosus, gracilaria sp., osmundea pinnatifida*), 3g/l fresh weight, to single-element and multi-element solutions (1μmol/l). of yttrium, lanthanum, cerium, praseodymium, neodymium, europium, gadolinium, terbium and dysprosium. There was a preference for light rare earths in single-element solutions, but this decreased when competing with other rare earths[62]. This competitive effect was less marked in the case of heavy rare earths. This indicated that these were still able to bind to the macro-algae in spite of the presence of competing ions. Unlike the water content, the seaweed's specific surface area was an important factor in the sorption of rare earths; a larger surface area was associated with greater removal and a larger competitive effect. In the specific case of europium (10, 152 or 500μg/l) in contaminated seawater, *ulva lactuca* and *gracilaria sp.* (3g/l fresh weight) proved[63] to be the most effective; attaining 85% europium removal within 72h. The highest europium enrichment was achieved by *ulva intestinalis* biomass; yielding up to 827μg/g. This is higher than the europium levels in most apatite ores. No cellular damage was suffered by the algae, but there were signs of toxicity and defence-mechanism activation. In the case of Nd^{III}, a macro-algae mass of 0.5, 3.0 or 5.5g/dm^3, a water salinity of 10, 20 or 30 and an initial Nd^{III} concentration of 10, 255 or 500μg/dm^3 was used[64]. Both algae revealed a clear ability to handle Nd^{III}; the removal being about 80% for an initial concentration of 255μg/dm^3, a salinity of 10 and a macro-algae mass of 5.5g/dm^3. Following 72h of exposure, the most important factor was the macro-algae mass. The higher the initial concentration, the greater was its accumulation.

The synthesis of algae/polyethyleneimine beads produces a stable absorbent for rare-earth elements[65]. The grafting-on of sulfonic groups which have a high affinity for rare earths increases the sorption capacities to as high as 2.68mmol/g for scandium, 0.61mmol/g for cerium and 0.53mmol/g for holmium. Sorption occurs within 30 to 40min, and the sorbent has a marked preference for Sc^{III} as compared with Ce^{III} and Ho^{III}. The sorbent is also selective for rare earths over alkali-earths. The above three metals are easily desorbed within 20 to 30min by using HCl/CaCl$_2$ solution. The de-sorption remains above 99% after 5 cycles, with the sorption performance decreasing by less than 6% at the fifth cycle.

Algal beads have been functionalized by phosphorylation and used for the sorption of Nd^{III} and Mo^{VI}. Phosphoryl groups were grafted as tributyl phosphate derivatives[66]. The multi-functional characteristics of the sorbent with regard to carboxylic, hydroxyl, amine and phosphate groups contributed to the binding of metal ions having differing physicochemical natures. The sorption of Nd^{III} was strongly increased by phosphorylation, but the increase in the case of Mo^{VI} was quite limited. Optimum sorption occurred for pH levels of 3 to 4, while the maximum sorption capacity attained 1.46mmol/g for Nd^{III} and 2.09mmol/g for Mo^{VI}. The sorption isotherms obeyed the Langmuir equation. Equilibrium was established within 30 to 40min and the kinetics were pseudo first-order. The effective diffusivity was close to the self-diffusivities of Nd^{III} and Mo^{VI} in water; reflecting the limited effect of any resistance to intraparticle diffusion upon the kinetics. The sorbent favoured Nd^{III} over Mo^{VI} or other metals at pH levels of 2.5 to 3. The metals were easily desorbed by using 0.2M HCl and 0.5M CaCl$_2$ as eluents. The loss of sorption capacity was less than 5% after 4 cycles, and desorption remained complete. Phosphorylation of algal/polyethylene-imine composite beads also markedly increases the sorption of La^{III} and Tb^{III}. Sorption occurs via interactions with phosphonate, amine and carboxylate groups. The sorption capacity at the optimum pH value of 5 attained 1.44mmol/g for lanthanum and 1.02mmol/g for terbium[67]. These were increases of 4.5 and 6.7 times[68], respectively, with respect to the figures for untreated beads. The loss of sorption capacity was less than 9%, and the desorption efficiency was greater than 99%. The sorbent exhibited a marked preference for rare earths over Si^{IV}, Ca^{II} and Mg^{II}. Functionalization also strongly increased the sorption efficiency of Pr^{III} and Tm^{III}. The optimum pH value was close to 5, then giving maximum sorption capacities that were close to 2.14mmol/g for praseodymium and 1.57mmol/g for thulium[69].

By using the symbiotic mixed culture, Kombucha, which consists of yeasts and acetic acid bacteria, rare earths have been leached at an appreciable rate[70]. The highest leaching-rates have been observed in cultures which use whole Kombucha as a leaching agent rather than its components, *zygosaccharomyces lentus* and *komagataeibacter hansenii*. During cultivation, the pH decreases as a result of acetic and gluconic acid production. In accord with the higher solubility of rare-earth oxides as compared to that of rare-earth phosphates and aluminates, the red dye $Y_2O_3:Eu^{2+}$ is found to be preferentially solubilized.

Various absorbents

Carbon-based nanomaterials

These carbonaceous materials contain functional groups which enhance the recovery of rare earths[71], and their complex structures permits surface modification for the sorption of rare earths. Graphene oxide is a two-dimensional material which possesses excellent mechanical properties combined with a high surface area and hydrophilic properties. Rare earths have a high affinity for the oxygen donors of the graphene oxide and can therefore be adsorbed at oxygen-based functional groups on the surface. On the other hand, in more complex aqueous environments grapheme oxide loses its efficiency. Various methods have been developed for graphene-surface modification in order to impart new properties to the material and thus improve sorption efficiency. Functionalization plays a pivotal role by controlling the material's surface and thus the pH used in processing. Among the ligands which have been used for rare earth recovery are ferrite and silica nanoparticles.

A methoxy-substituted tripodal hydroxylamine ligand, H_3TriNO_xOMe, has been coordinated[72] to rare-earth cations for separation purposes. The addition of an electron-donating group to the aryl backbone led to a more electron-rich ligand that increased, by

5 times, the equilibrium constant for complex dimerization. This separation system provided efficient neodymium/dysprosium separation in toluene.

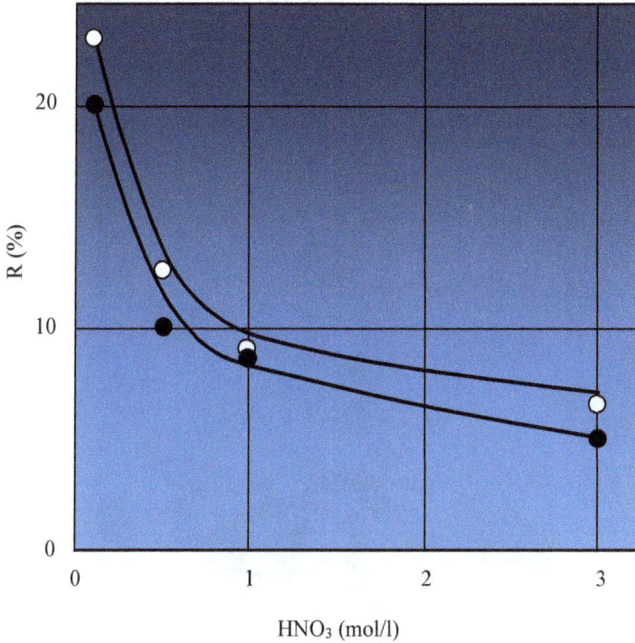

Figure 1. Sorption extraction of rare earths, in nitric acid solution, by oxidized multilayer graphene

Ethylene di-aminetetra-acetic acid, diethylenetri-aminepenta-acetic acid, diglycolamic acid and humic acids have been applied to carbon-based nanomaterials. Europium, cerium, gadolinium, scandium, yttrium, lanthanum and neodymium have been recovered from e-waste by using both batch and column methods at room temperature. The use of nanocomposites is effective in the recovery of rare earths in elemental form because their adsorption depends upon the pH level.

Table 15. Degree of extraction of rare earths from solutions
(3mol/l HNO₃) by sorption on modified oxidized multilayer graphene

Oxidized Multilayer Graphene Ligand	Rare Earth	Extraction (%)
none	Ce	3
none	La	3
none	Eu	4
diphenyl-dibutyl-carbamoyl-methyl-phosphine oxide	Ce	90
diphenyl-dibutyl-carbamoyl-methyl-phosphine oxide	La	89
diphenyl-dibutyl-carbamoyl-methyl-phosphine oxide	Eu	91
tetra-octyl-diglicolamide	Ce	93
tetra-octyl-diglicolamide	La	92
tetra-octyl-diglicolamide	Eu	99
trioctyl-phosphinoxide	Ce	2
trioctyl-phosphinoxide	La	0
trioctyl-phosphinoxide	Eu	6
di-(2-ethylhexyl)-phosphoric acid	Ce	8
di-(2-ethylhexyl)-phosphoric acid	La	4
tributyl phosphate	Ce	8
tributyl phosphate	La	5
tributyl phosphate	Eu	8

A polyurethane-sponge supported titanium phosphate with graphene oxide, prepared by *in situ* precipitation, has been tested[73] for the recovery of traces of dysprosium from water by using batch experiments. The sponge exhibited a strong affinity for dysprosium, with the theoretical capacity attaining 576.17mg/g and half-equilibrium being reached within 2.5min. The sponge also exhibited adsorption over a wide range of pH-values and salinity. Its behaviour was explained in terms of a strong binding of the phosphate to Dy^{III}, an enhanced surface area due to the presence of graphene oxide and lower aggregation of the spongy structure. The main adsorption mechanism involved electrostatic interaction. Graphene oxide has also been carboxylated, and modified to give hydroxylated graphene oxide[74]. Diatomaceous earth and chitosan were loaded by solution blending. Carboxylated graphene oxide – diatomite chitosan and carboxylated graphene oxide diatomite magnetic chitosan composites were prepared via simple solid–liquid separation. These modified composites could be used to remove lanthanum. The adsorption effect was optimum for an initial solution concentration of 50mg/g, a pH-value of 8.0, 3g/l of adsorbent and a temperature and adsorption time of 45C and 50min. The adsorption process was consistent with a pseudo second-order kinetic model and the Langmuir model, and internal diffusion was not the only governing effect. The adsorption process was endothermic. The maximum adsorption capacity of the carboxylated graphene oxide diatomite magnetic chitosan for La^{III} at 308K was 302.51mg/g. Following 4 adsorption-desorption cycles, the adsorption capacity of this composite initially exceeded 74%.

Mesoporous magnetic microsphere rare-earth adsorbents in the form of MnFe$_2$O$_4$-Al$_2$O$_3$ and MnFe$_2$O$_4$|SiO$_2$-chitosan have been prepared[75], with MnFe$_2$O$_4$ spinel being the main component. Under the optimum condition of a pH of 7.0 at 298K, the maximum adsorption capacities of MnFe$_2$O$_4$|SiO$_2$-chitosan with regard to La^{3+} and Ce^{3+} were 1030 and 1020mg/g, respectively. The adsorption reactions could reach equilibrium within 0.5h. The adsorbents could also be reused. The adsorption of chitosan/nano-SiO$_2$ has been studied[76] in connection with low concentrations of rare earth ions, showing that the optimum conditions were a temperature of 25C, a pH level of 5 and an initial mass concentration of 45, 37.5 or 27.5mg/l in the case of gadolinium, lanthanum or yttrium, respectively. The absorbent dosage was 40mg. Under these conditions, the saturated adsorption capacity was 22.3, 17.8 and 12.9mg/g for gadolinium, lanthanum and yttrium, respectively. The Langmuir isotherm could be used to describe the adsorption of the rare earth ions, and experimental results showed that the chitosan/nano-SiO$_2$ interacted strongly with the rare earth ions, leading to an adsorption efficiency greater than 98%. The rare earth ions could be desorbed by using hydrochloric acid.

The sorption behavior of oxidized multilayer graphene with respect to actinides and rare earths in nitric acid (< 3mol/l) solutions was studied[77]. The oxidized multilayer graphene was modified by using tetra-octyldiglycolamide, diphenyl-dibutyl-carbamoyl-methyl-phosphine oxide, tri-octyl-phosphinoxide, di-(2-ethylhexyl) phosphoric acid, tributyl phosphate and di-2-ethyl-hexyl-methyl-phosphonate reagents. The formation time of solid-phase compacts was 20 to 240min. The sorption capacities with regard to U^{VI}, Th^{IV}, Pu^{IV}, La^{III}, Ce^{III} and Eu^{III} in nitric acid solutions (3mol/l) were determined (figure 1, table 16). The distribution coefficients of the elements ranged from 10^3 to 10^{4}ml/g.

Carbon nanotubes are one-dimensional nanocomposites and can also be used for the recovery of rare earths (table 17). Carbon nanotube composites are effective in the recovery of rare earths by using distilled water. The sorption of rare earths when using this process depends upon the residence-time, pH level and temperature. The recovery of rare earths is also affected by the type of carbon nanotube, with oxidized multi-walled nanotubes being cheaper and more efficient than single-walled carbon nanotubes. Carbon nanotubes can be divided into two main groups. Single-walled carbon nanotubes are effectively a single sheet of graphene which has been rolled up to form a cylindrical tube, while multi-walled carbon nanotubes consist of a set of concentric nanotubes which are stabilized by van der Waals forces. The presence of concentric graphene sheets in the latter enhances interactions. The single-walled carbon nanotubes can have three distinct forms: armchair, zig-zag or chiral. An important property is their insolubility in water and in almost all other solvents. In order to disperse nanotubes in liquids, functional groups or polar molecules can be incorporated without greatly affecting their properties.

Table 16. Recovery of rare earths by using graphene oxide composites

Sorbent	pH	T (C)	Contact Time (h)	Rare Earth	Recovery (mg/g)
GO colloid	6	25	0.5	lanthanum	85.7
GO colloid	6	25	0.5	neodymium	189
GO colloid	6	25	0.5	gadolinium	226
GO colloid	6	25	0.5	yttrium	136
GO	5.5	20	0 to 24	europium	143
GO-OSO$_3$H	5.5	20	0 to 24	europium	125
GO colloid	2-11	30	0.5	gadolinium	287
GO colloid	5.9	30, 40	0.42	yttrium	190
GO	2.7-7.3	25	48	europium	78.0
GO	2	25	4	scandium	36.5
GO	4	25	4	scandium	39.7
MPANI-GO	4	25	0.33	yttrium	8.10
MPANI-GO	4	25	0.33	lanthanum	15.5
MPANI-GO	4	25	0.33	cerium	8.60
MPANI-GO	4	25	0.33	praseodymium	11.1

MPANI-GO	4	25	0.33	neodymium	8.50
MPANI-GO	4	25	0.33	samarium	7.70
MPANI-GO	4	25	0.33	europium	11.0
MPANI-GO	4	25	0.33	gadolinium	16.3
MPANI-GO	4	25	0.33	terbium	11.8
MPANI-GO	4	25	0.33	dysprosium	16.0
MPANI-GO	4	25	0.33	holmium	8.10
MPANI-GO	4	25	0.33	erbium	15.2
MPANI-GO	4	25	0.33	terbium	10.4
MPANI-GO	4	25	0.33	ytterbium	10.3
MPANI-GO	4	25	0.33	lutetium	14.9
PANI|GO	3	25	48	europium	251

The maximum adsorption of rare earths by carbon nanotube composites depends greatly upon the pH level because this affects the surface charge and thus the sorption of metal ions. Increasing the pH level generally increases metal-ion sorption because, at pH levels higher than the point of zero charge, the positively-charged metal ions can be adsorbed on negatively-charged oxidized carbon nanotubes. The most commonly chosen pH level is 5, although values of 1.5 and 8 have been used.

Table 17. Recovery of rare earths by using carbon nanotubes

Sorbent	pH	T (C)	Contact Time (h)	Rare Earth	Recovery (mg/g)
CNT-COOH	2	25	4	scandium	37.9
CNT-COOH	4	25	4	scandium	42.5
TA-MWCNT	5	20	1	lanthanum	5.35
TA-MWCNT	5	20	1	terbium	8.55
TA-MWCNT	5	20	1	lutetium	3.97
mIIP-CS/CNT	7	20	4	gadolinium	79.5
mIIP-CS/CNT	7	33	4	gadolinium	109
mIIP-CS/CNT	7	43	4	gadolinium	122
mNIP-CS/CNT	7	33	4	gadolinium	96.2

Unaligned carbon nanotubes, encapsulated in polyvinyl alcohol, have been used[78] to produce an electrochemically active filter for the recovery of copper, arsenic, europium, neodymium, gallium and scandium. At flow-rates of 1 to 5ml/min, pH levels of 2 to 10 and voltages of 0.1 to 3.0V, the maximum recovery rates were 86 to 96%; except in the case of arsenic, which was recovered. All of the metals were generally recovered as oxides, apart from copper, which was partially reduced at low pH levels. De-aeration studies suggested that the electrochemical reduction of dissolved O_2, and of O_2 resulting from water-splitting, were together responsible for metal-capture. Metal oxides were initially formed via metal-hydroxide intermediaries; a mechanism which was enhanced by higher pH levels. A waste stream of copper and europium could be separated in several stages at increasing voltages, leading to 97% copper recovery and 65% europium recovery.

Activated carbon (tables 18 to 20) is very porous and its principal characteristics thus include a high internal surface area and innumerable internal spaces which can be classified into micro-, meso- and macro-.

*Table 18. Recovery of rare earths using functionalized activated carbon
(ultrapure water, neutral pH, room-temperature, 24h contact time)*

Sorbent Content (mg/l)	Rare Earth	Adsorption (%)
0.25×10^2	cerium	12
0.15×10^2	cerium	11
0.05×10^2	cerium	8.0
0.03×10^2	cerium	2.5
0.25×10^2	lanthanum	7.5
0.15×10^2	lanthanum	6.5
0.05×10^2	lanthanum	2.5
0.03×10^2	lanthanum	1.5
0.25×10^2	neodymium	31
0.15×10^2	neodymium	24
0.05×10^2	neodymium	17
0.03×10^2	neodymium	9.0
0.25×10^2	samarium	7.5
0.05×10^2	samarium	7.5
0.15×10^2	samarium	5
0.03×10^2	samarium	0
0.25×10^2	yttrium	12.5
0.15×10^2	yttrium	11
0.05×10^2	yttrium	9.0
0.03×10^2	yttrium	6.0

Materials Research Forum LLC

https://doi.org/10.21741/9781644901793

Table 19. Recovery of rare earths using activated carbon
(Milli-Q water, neutral pH)

Sorbent Content (mg/l)	T (C)	Contact Time (h)	Rare Earth	Adsorption (%)
0.5×10^2	80	1	cerium	95
0.5×10^2	80	2	cerium	95
0.03×10^2	25	24	cerium	1.0
0.05×10^2	25	24	cerium	1.0
0.5×10^2	80	2	lanthanum	45
0.5×10^2	80	1	lanthanum	40
0.05×10^2	25	24	lanthanum	1.5
0.03×10^2	25	24	lanthanum	1.0
0.5×10^2	80	2	neodymium	80
0.5×10^2	80	1	neodymium	75
0.05×10^2	25	24	neodymium	8.0
0.03×10^2	25	24	neodymium	7.5
0.5×10^2	80	2	samarium	82
0.5×10^2	80	1	samarium	80
0.05×10^2	25	24	samarium	1.0
0.03×10^2	25	24	samarium	0
0.5×10^2	80	2	yttrium	72
0.5×10^2	80	1	yttrium	63
0.03×10^2	25	24	yttrium	1.5
0.05×10^2	25	24	yttrium	1.0

Table 20. Recovery of europium using HPO_4-APC activated carbo
(ultrapure water, pH = 5, 20C, room-temperature, 2h contact time)

Sorbent Content (mg/l)	Adsorption (%)
17.5×10^2	93
15×10^2	90
12.5×10^2	80
10×10^2	72
5×10^2	60
7.5×10^2	60
2.5×10^2	45

Ethylenediaminetriacetic acid functionalized activated carbon has been synthesized[79] by anchoring N-[(3-trimethoxysilyl)propyl]ethylenediaminetriacetic acid to oxidized activated carbon. The maximum rare-earth adsorption capacity for neodymium was deduced by constructing an adsorption isotherm and fitting the data to the Langmuir adsorption model. The material's affinity for each lanthanide ion was determined and this showed that, among binary mixtures of La/Ni, Sm/Co, Eu/Y and Dy/Nd, the highest selectivity was observed for heavy rare earths. The adsorbed metal ions could be recovered, and the adsorbent could be regenerated, by treatment with dilute hydrochloric acid. Modified activated carbon has also been synthesized[80] by loading it with pentaethylenehexamine. Plain and modified activated carbon samples were contacted with lanthanum solutions, and the latter's adsorption was determined. The above modification increased adsorption from 44 to 100% and increased release from 65 to 91%, with respect to plain activated carbon; giving an overall recovery efficiency of 90%. Magnetic ordered mesoporous carbon has been used[81] as a core for the preparation of imprinted material, using gadolinium as a template. The prepared material was then used as a coated sorbent for solid-phase micro-extraction fiber, and was also introduced into a micropipette tip in order to perform microsolid phase extraction. The fiber exhibited a pre-concentration factor of 1400 for gadolinium, with a detection limit of 2.34ng/l, whereas the microsolid phase extraction offered an adsorption capacity of 30.2µg/g and a gadolinium removal efficiency of 90%. Both techniques could remove gadolinium from a wide range of waste-waters.

A carbon nanofiber is a non-continuous 1-dimensional nano-allotrope of cylindrical or conical form which consists of stacked curved graphene sheets. It can be described as a sp^2-based linear filament having a diameter of 50 to 200nm and an aspect ratio greater than 100. The surface properties can be modified by chemical treatment, so as to create an effective adsorbent.

A fullerene is a molecule of carbon in the form of a hollow sphere, tube or other shape. It is essentially a closed hollow cage which is made up of sp^2-hybridized carbon atoms arranged into 12 pentagons, plus enough hexagons to satisfy geometrical constraints and leading to C_{60} being the most abundant form.

Carbon dots are quasi-spherical carbon nanoparticles, with a diameter of 2 to 10nm, which have a high oxygen content and comprise combinations of graphitic and turbostratic carbon in various ratios. Their most significant property in the present context is surface functionalization.

Carbon black (tables 21 to 23) is produced by the incomplete combustion of heavy petroleum products plus a small amount of vegetable oil. It is a form of paracrystalline carbon that possesses a high ratio of surface-area to volume; albeit lower than that of activated carbon.

Table 21. Recovery of rare earths using functionalized commercial carbon black (ultrapure water, neutral pH, room temperature, 24h contact time)

Sorbent Content (mg/l)	Rare Earth	Adsorption (%)
0.25×10^2	cerium	41
0.03×10^2	cerium	36
0.05×10^2	cerium	36
0.15×10^2	cerium	35
0.25×10^2	lanthanum	15
0.15×10^2	lanthanum	14
0.05×10^2	lanthanum	13
0.03×10^2	lanthanum	12
0.25×10^2	neodymium	23
0.15×10^2	neodymium	16
0.03×10^2	neodymium	12

0.05×10^2	neodymium	12
0.25×10^2	samarium	14
0.15×10^2	samarium	13
0.03×10^2	samarium	10
0.05×10^2	samarium	10
0.25×10^2	yttrium	17
0.05×10^2	yttrium	13
0.15×10^2	yttrium	13
0.03×10^2	yttrium	12

Table 22. Recovery of rare earths using commercial carbon black (ultrapure water, neutral pH, room temperature, 24h contact time)

Sorbent Content (mg/l)	Rare Earth	Adsorption (%)
0.15×10^2	cerium	1.0
0.25×10^2	cerium	1.0
0.15×10^2	lanthanum	2.5
0.25×10^2	lanthanum	2.5
0.25×10^2	neodymium	8.0
0.15×10^2	neodymium	5.0
0.25×10^2	samarium	2.5
0.15×10^2	samarium	1.0
0.25×10^2	yttrium	3.0
0.15×10^2	yttrium	2.5

Materials Research Forum LLC
https://doi.org/10.21741/9781644901793

Table 23. Recovery of rare earths using recycled-tyre carbon black
(ultrapure water, neutral pH)

Sorbent Content (mg/l)	T (C)	Contact Time (h)	Rare Earth	Adsorption (%)
0.5×10^2	80	12	cerium	95
0.5×10^2	25	12	cerium	95
0.5×10^2	25	2	cerium	90
0.5×10^2	60	24	cerium	90
0.5×10^2	80	24	cerium	90
0.5×10^2	25	1	cerium	85
0.5×10^2	40	24	cerium	85
0.25×10^2	80	24	cerium	84
0.25×10^2	60	24	cerium	81
0.25×10^2	40	24	cerium	75
0.25×10^2	25	24	cerium	68
0.15×10^2	25	24	cerium	42
0.05×10^2	80	24	cerium	30
0.05×10^2	60	24	cerium	25
0.05×10^2	40	24	cerium	23
0.05×10^2	25	24	cerium	15
0.03×10^2	25	24	cerium	11
0.5×10^2	80	12	lanthanum	75
0.5×10^2	80	24	lanthanum	69
0.5×10^2	25	12	lanthanum	60
0.5×10^2	60	24	lanthanum	52
0.25×10^2	80	24	lanthanum	48
0.5×10^2	25	2	lanthanum	45
0.5×10^2	40	24	lanthanum	45

0.25×10^2	60	24	lanthanum	32
0.25×10^2	40	24	lanthanum	29
0.25×10^2	25	24	lanthanum	28
0.5×10^2	25	1	lanthanum	25
0.15×10^2	25	24	lanthanum	18
0.05×10^2	80	24	lanthanum	13
0.05×10^2	60	24	lanthanum	7.5
0.05×10^2	25	24	lanthanum	6.0
0.05×10^2	40	24	lanthanum	5.5
0.03×10^2	25	24	lanthanum	3.5
0.5×10^2	80	12	neodymium	91
0.5×10^2	25	12	neodymium	83
0.5×10^2	80	24	neodymium	75
0.5×10^2	25	2	neodymium	70
0.5×10^2	60	24	neodymium	70
0.5×10^2	25	1	neodymium	68
0.5×10^2	40	24	neodymium	65
0.25×10^2	80	24	neodymium	58
0.25×10^2	60	24	neodymium	50
0.25×10^2	40	24	neodymium	40
0.25×10^2	25	24	neodymium	34
0.15×10^2	25	24	neodymium	22
0.05×10^2	80	24	neodymium	20
0.05×10^2	60	24	neodymium	16
0.05×10^2	40	24	neodymium	9.0
0.05×10^2	25	24	neodymium	7.5
0.03×10^2	25	24	neodymium	5.0

0.5×10^2	80	12	samarium	95
0.5×10^2	25	12	samarium	88
0.5×10^2	80	24	samarium	75
0.5×10^2	25	2	samarium	73
0.5×10^2	60	24	samarium	72
0.5×10^2	40	24	samarium	68
0.5×10^2	25	1	samarium	60
0.25×10^2	80	24	samarium	60
0.25×10^2	60	24	samarium	55
0.25×10^2	25	24	samarium	41
0.25×10^2	40	24	samarium	40
0.15×10^2	25	24	samarium	26
0.05×10^2	80	24	samarium	20
0.05×10^2	60	24	samarium	16
0.05×10^2	25	24	samarium	9.0
0.05×10^2	40	24	samarium	9.0
0.03×10^2	25	24	samarium	5.5
0.5×10^2	80	12	yttrium	90
0.5×10^2	25	12	yttrium	77
0.5×10^2	80	24	yttrium	75
0.5×10^2	60	24	yttrium	70
0.5×10^2	25	2	yttrium	60
0.25×10^2	80	24	yttrium	60
0.5×10^2	40	24	yttrium	60
0.25×10^2	60	24	yttrium	50
0.5×10^2	25	1	yttrium	48
0.25×10^2	40	24	yttrium	40

0.25×10^2	25	24	yttrium	28
0.05×10^2	80	24	yttrium	21
0.15×10^2	25	24	yttrium	18
0.05×10^2	60	24	yttrium	16
0.05×10^2	40	24	yttrium	9.0
0.05×10^2	25	24	yttrium	6.0
0.03×10^2	25	24	yttrium	3.5

The material having the highest maximum adsorption capacity for rare earths appears to be oxygen- and phosphorus-functionalized nanoporous carbon, with values of 335.5 and 344.6mg/g being observed for neodymium and dysprosium, respectively.

Silica

The composition of the immobilized layer plays a critical role in the metal-adsorption ability of complexing organo-mineral materials. It has been demonstrated[82] that suitable surface-assembled synthesis of organo-silica with covalently immobilized fragments of dipicolinic acid results in an adsorbent that is capable of recovering almost all rare earths from multi-element solutions having pH levels greater than 1.7. No noticeable loss of efficiency was found, and the mean degree of rare-earth recovery was greater than 97%. The adsorbent was used to recover rare earths from model solutions which consisted of 22 metal ions in a 0.5mol/l NaCl solution. Even a 3200-fold excess of iron and copper ions only slightly impaired the rare-earth recovery. The adsorbent was able to recover more than 80% of all rare earths from the acidic leachants of fluorescent lamps, with enrichment factors greater than 600. Lanthanum was an exception. Following the adsorption of Eu^{3+} and Tb^{3+}, the materials exhibited a strong red or green luminescence, respectively. This indicated the operation of a chelating mechanism in rare-earth adsorption on silica with dipicolinic acid. Hybrid adsorbents have been produced via the surface modification, with amino polycarboxylate ligands, of mesoporous nanostructured silica beads[83]. Cytotoxicity was assessed by using human muscle-derived cells, fibroblast cells, macrophage cells and human umbilical vein endothelial cells. This indicated a lower toxicity of ligand-free materials than that of materials with amino poly-carboxylate functionalization. Cell-internalization of the material, and the release of nitric oxide, were observed. Zebrafish embryos which were exposed to high concentrations of the material did not exhibit any pronounced toxicity. New sorbents have been developed[84] for the recovery of rare-earth ions from aqueous solutions by grafting diethylenetriamine onto silica-composite supports. The sorption capacity at the optimum pH value of 4 attained 0.9mmol/g of neodymium and 1mmol/g of gadolinium. At pH levels of 4 to 5, the sorbent exhibits a high selectivity for rare earths with respect to alkali earths. Selectivity was confirmed by the efficient recovery of rare earths from acidic leachates of gibbsite. Following selective precipitation using oxalate solutions, and calcination, pure rare-earth oxides are obtained. Absorbents having ethylenediaminetriacetic, phosphonic and ammonium groups as ligands have been synthesized[85] by using the template method, and sodium metasilicate as the main starting reagent. They have an ordered mesoporous structure, with a uniform pore diameter and a large surface area. The adsorption mechanisms involve mainly complexation and electrostatic interactions between metal ions and the various functional groups on the surface. The synthesized materials can remove Fe^{III}, Ni^{II}, Cu^{II}, Pb^{II}, Nd^{III} and Dy^{III} from water with a pH greater than 1. The highest adsorption capacity was found for bifunctional EDTA/phosphonic-modified material, and attained up to 119.05mg/g for Fe^{III}, 246.95mg/g for Pb^{II}, 246.95mg/g for Cu^{II}, 238.10mg/g for Nd^{III} and 243.90mg/g for Dy^{III}.

Pyrometallurgy

Pyrometallurgy is a familiar thermal treatment for the recovery of metals from electronic waste and involves melting in a blast furnace or plasma arc furnace and heating in the presence of suitable gases in order to recover mainly non-ferrous metals. It offers high-efficiency recovery, of the order of 70%, of certain rare earths from e-waste. The smelting step can accept any e-waste feedstock for the recovery of copper and precious metals from printed-circuit boards. Treatment of the latter yields a mixed oxide slag of mainly lead and zinc, together with a Cu-Ni-Si alloy. The use of high pressures improves the separation of antimony, bismuth, lead and other heavy metals. Vacuum pyrometallurgy can recover rare earths from e-waste. Attention tends to be focused on waste products such as high-coercivity magnets and nickel-metal-hydride batteries. The processes can include electro-slag re-melting, extraction by using molten magnesium, high-temperature treatment with metal halides and chlorides and the electrolysis of molten salts[86]. A study was made[87] of the phase relationships in the $CaO\text{-}SiO_2\text{-}Nd_2O_3$ system with regard to the high-temperature recycling of neodymium. Slag samples were equilibrated at 1500C and 1600C for 24h under argon, and quenched in water. On the basis of the established phase relationships, a solidification process involving various cooling paths could be identified for the purpose of recycling.

When rare earths, indium, cobalt, lithium, gold silver and platinum were recovered[88] from electronic components by pyrolysis the process also produced liquid and gaseous mixtures of organic compounds that had some value as fuels, but also generated particulate matter and semi-volatile organic products while the ash residue contained leachable pollutants such as arsenic, chromium, cadmium and lead. This suggests that rare-earth recovery processes themselves carry some environmental risks.

Electrochemistry

Electrochemical processes can be used for the extraction or enrichment of metals from e-waste. When used for extraction, scrap material can be made the anode and the application of electromotive force drives chemical reactions. During this process, water is oxidized at the anode to produce hydrogen. Proton accumulation at the anode decreases the pH level and dissolves metals. Electrons flow towards the cathode via an external circuit and electrolysis occurs, thus generating hydrogen gas and hydroxide ions. Charge-balance is maintained by anions and cations migrating between chambers, leading to a solution which contains a mixture of rare earths ready for further separation. The overall process can be described by:

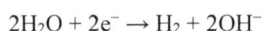

$$2H_2O \rightarrow 4H^+ + O_2 + 4e^-$$

$$2H_2O + 2e^- \rightarrow H_2 + 2OH^-$$

$$2H^+ + 2e^- \rightarrow H_2$$

Various side-reactions can occur, and the deposition of iron, nickel, cobalt, etc., can occur; depending upon the type of scrap and the cathode composition. Electrochemical processing can also be used as a complementary treatment during rare-earth purification. The recovery of rare earths can be improved by increasing the voltage in an electrochemical filtration system. The method has been most widely applied to rare earth recovery from Nd-Fe-B magnets, and a marked advantage is the lower consumption of chemicals when compared with hydrometallurgy.

An electrochemical recovery process which is based upon the regeneration of ferric ions has been studied[89] for the selective recovery of base metals, while separate processes recovered precious metals and rare earths from magnets. Recovery and extraction efficiencies of about 90% were found for the extraction of base metals from the non-ferromagnetic fraction in solutions of sulfuric and hydrochloric acid. The extraction of rare earths from the ferromagnetic fraction was achieved by anaerobic extraction in acid media. A study[90] of the properties of choline-based ionic liquids showed that the electrical conductivity depended upon the density of the liquids, and that rare earth species affect the conductive mechanism. This was relevant to the recycling of rare earths from Nd-Fe-B magnets by electrodeposition using ionic liquids. The recycling process involved de-magnetization, chemical etching, dissolution, synthesis of metallic salts and two-stage electrodeposition. The results proved that iron-group and rare-earth metals can be separately recovered by electrodeposition.

An electrochemical method involving molten salts has been considered[91] for the recycling of nickel mischmetal hydride batteries. The electrochemical behaviour of lanthanum, cerium and neodymium was investigated in molten LiCl-KCl eutectic, containing LiF, by using tungsten and nickel electrodes. When using tungsten, the electro-reduction peaks of each rare earth in cyclic voltammograms became clearer with increasing LiF content. When using a nickel electrode, no marked effect of fluoride additions upon the potential for the electro-reduction of rare earths could be identified due to the alloying of rare earths with nickel. It was concluded that, in order to co-extract rare earths from a molten bath, active electrodes were suitable but, in order to separate neodymium from other rare earths, even inert electrodes were suitable for species control in the electrolytes.

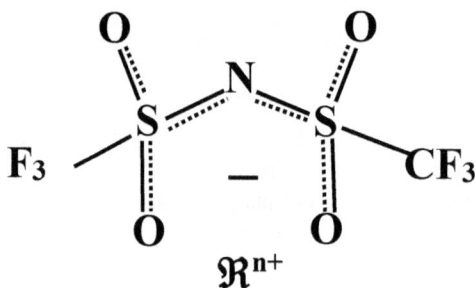

Figure 2. Structure of metallic-bis(trifluoromethyl-sulfonyl)amide, n = 1 to 3

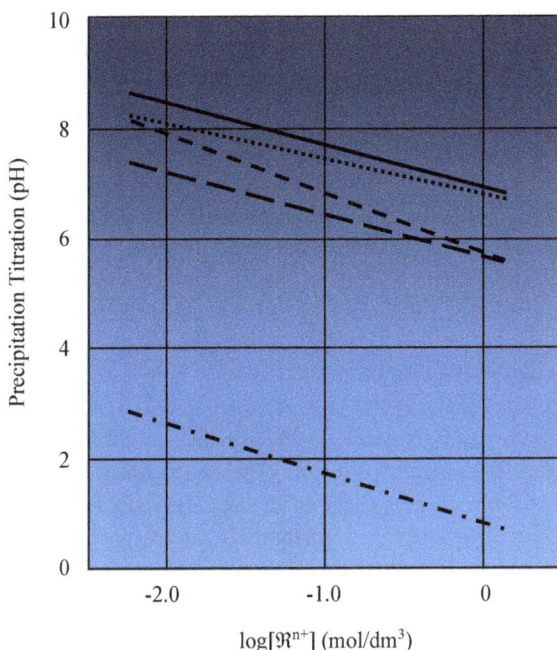

Figure 3. $\mathfrak{R}^{+}$ precipitation titration in bis(trifluoromethyl-sulfonyl)amide solution. Solid line: Pr^{3+}, dotted line: Nd^{3+}, short-dash line: Dy^{3+}, long-dash line: Fe^{2+}, dash-dot line: Fe^{3+}.

The electrodeposition of neodymium and praseodymium from molten NdF_3 + PrF_3 + LiF + $1wt\%Pr_6O_{11}$ + $1wt\%Nd_2O_3$ and NdF_3 + PrF_3 + LiF + $2wt\%Pr_6O_{11}$ + $2wt\%Nd_2O_3$ electrolytes at 1323K was investigated[92], showing that a critical condition for co-deposition was a constant praseodymium deposition over-potential of about −0.100V. This led to co-deposition current densities of the order of $6mA/cm^2$. Praseodymium deposition occurred via a 1-step process which involved the exchange of 3 electrons Pr^{III} → Pr^0. Neodymium deposition was a 2-step process which involved 1-electron exchange, Nd^{III} → Nd^{II}, followed by the exchange of 2 electrons, Nd^{II} → Nd^0. Diffraction analysis confirmed the formation of metallic neodymium and praseodymium on the substrate.

It has been found[93] that $NdCl_3$ is advantageous, as a general precursor to room-temperature neodymium electrodeposition, due to its abundance and availability. It has however poor solubility in most specialized ionic liquids and is not widely used in solvometallurgical processes. A means of neodymium electrodeposition from a solution of $NdCl_3$ and $AlCl_3$ in 1,3-dimethyl-2-imidazolidinone has however been found. Aluminium electrodeposition did not occur at the $AlCl_3$ concentrations used.

The concentration dependence of the precipitation reaction of alkali metal hydroxide against Fe^{2+}, Fe^{3+} and $\mathfrak{R}^{3+}$ ($\mathfrak{R}$ = praseodymium, neodymium, dysprosium) was investigated[94], showing that $Fe(OH)_3$ precipitates were easy to form in media at acidic pH levels (figures 2 and 3). This was then exploited as a wet separation method for the recovery of rare earths from Nd-Fe-B magnet scrap, combined with demagnetization, chemical etching, roasting, acid-leaching, precipitation titration, synthesis of metal-bis(trifluoromethyl-sulfonyl) amide salts and electrodeposition. Following precipitation titration, metal-bis(trifluoromethyl-sulfonyl) amide salts were synthesized in yields greater than 90% by evaporation. A clear endothermic peak of the amide salts appeared, and the melting point was close to that of neodymium-[bis(trifluoromethyl-sulfonyl)]₃ and potassium-bis(trifluoromethyl-sulfonyl). The best method was to use these metal-bis(trifluoromethyl-sulfonyl) amide salts as an electrolyte bath for electrodeposition. The electrodeposition of neodymium was performed under a potential of -3.5V, versus a platinum quasi-reference electrode, at 513K. The electrodeposits were neodymium metal in their middle layers. The carbon content was extremely low and the oxygen content steadily decreased with depth of the electrodeposits.

Hydrometallurgy

This is a chemical method which can extract metals such as gadolinium, yttrium and lanthanum from e-waste by using two steps. In the first step, metals are leached from the waste by using acids such as sulfuric, nitric and hydrochloric. In the second step, the dissolved metals are recovered from the leachate by using adsorption, liquid-liquid extraction, cementation precipitation or electro-winning. One of earliest processes[95] to be used to separate rare earths from Nd-Fe-B magnet scrap involved sulfuric acid dissolution, followed by the precipitation of rare-earth salts. Resultant sodium and ammonium double-salt precipitates could be converted into various neodymium products. In the case of sodium salts, the neodymium recovery-rate was 98%. In the case of ammonium salts it was 70%. Iron-rich effluent could be treated to produce sodium- and ammonium-iron Jarosites that could be converted to hematite. This method also avoided the disadvantages of using fluoride or oxalate precipitation.

Table 24. Recovery of rare earths by hydrometallurgy

Leachant	Metal	Efficiency (%)
2M H_2SO_4	cerium	35
$H_2O + H_2SO_4$	cerium	98.8
HNO_3	dysprosium	81
HNO_3, HCl or H_2SO_4	europium	>90
4M ($HNO_3 + H_2SO_4$)	europium	92.8
2M H_2SO_4	lanthanum	35
$H_2O + H_2SO_4$	lanthanum	80.4
H_2SO_4	neodymium	95-99
HNO_3	neodymium	98
$H_2O + H_2SO_4$	neodymium	98.2
$H_2O + H_2SO_4$	praseodymium	98.5
$H_2O + H_2SO_4$	samarium	99.2
HNO_3, HCl or H_2SO_4	yttrium	>90
H_2SO_4	yttrium	85
4M ($HNO_3 + H_2SO_4$)	yttrium	96.4
H_2SO_4	yttrium	99

Rare earths can be extracted from e-waste by using super-critical water, and leaching with hydrochloric acid. This unfortunately requires temperatures of 420 to 440C. An alternative is to use supercritical CO_2 combined with a mixture of sulfuric acid and hydrogen peroxide. Fluid extraction has been used[96] to recover rare earths from waste fluorescent lamps by means of super-critical CO_2 solvent, together with tributyl-phosphate nitric acid chelating agent. Rare-earth extraction efficiencies of 50% were possible in the absence of sample pre-treatment. Ball-milling for 1h led to a 20% improvement in extraction efficiency. During this mechanical activation, the sample became nano-crystalline and this led to an increased leaching efficiency. The mechanical activation of waste fluorescent powders can also be achieved by using a vibratory-disc mill[97]. The mechanical forces which are imposed on the powders produce crystal structural defects and thereby increase the rare-earth leachability. For example, preliminary activation increased terbium dissolution by 35%. Subsequent precipitation and calcination then led to a purity of 98.3%, as oxide. Magnetic levitation can be used[98] to separate heavy metals from metallic mixtures. The technique is able to levitate substances having densities greater than 5.00g/cm³. It was used to separate rare earths from fluorescent powders, as well as indium from indium-tin oxide and glass mixtures such as mechanically processed liquid crystal displays.

Temperature plays an important role in leaching rare earths from spent fluorescent lamps. The use of 4M H_2SO_4 at 90C can leach more than 95% of the yttrium from such lamps. The H_2SO_4 leaching of europium and yttrium from their oxides in the lamps occurs according to:

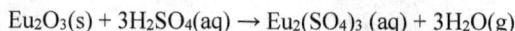

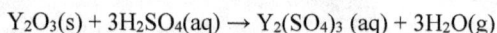

$$Eu_2O_3(s) + 3H_2SO_4(aq) \rightarrow Eu_2(SO_4)_3 \,(aq) + 3H_2O(g)$$

$$Y_2O_3(s) + 3H_2SO_4(aq) \rightarrow Y_2(SO_4)_3 \,(aq) + 3H_2O(g)$$

The term, thermomorphic, means that the solubility of an ionic liquid in water can be changed by temperature and it can even become immiscible. It is possible that this property could be used for the leaching and extraction of yttrium and europium from scrap fluorescent lamp phosphors.

Tri-colour phosphors are coated as thin layers onto the insides of fluorescent lamps for the purpose of converting invisible ultra-violet radiation into visible light. The phosphors are of four types: phosphate, aluminate, borate and silicate, each involving red-, green- and blue-generating materials. Regardless of the system used, the red phosphor is always composed of Y_2O_3:Eu^{3+}. In the phosphate system, the green phosphor is composed of $LaPO_4$:Ce^{3+},Tb^{3+} and the blue phosphor is composed of $(Ba,Sr,Ca)_5(PO_4)_3Cl$:Eu^{2+}. In the aluminate system, the green phosphor is composed of $CeMgAl_{11}O_{19}$:Tb^{3+} and the blue phosphor is composed of $BaMgAl_{10}O_{17}$:Eu^{2+}. In the borate system, the green phosphor is composed of $GdMgB_5O_{10}$:Ce^{3+},Tb^{3+} and the blue phosphor is composed of $Ca_2B_5O_8Cl$:Eu^{2+}. In the silicate system, the green phosphor is composed of Y_2SiO_3:Ce^{3+},Tb^{3+} and the blue phosphor is composed of $BaZrSi_3O_9$:Eu^{2+}. The aluminate system is the one which is most widely used, and thus aluminates are the commonest targets for recycling. The task is therefore to treat the various oxides which are present. The red phosphors consist of 85.3 to 93.0% of Y_2O_3 and 6.8 to 7.6% of Eu_2O_3. The green phosphors consist of 59.1 to 72.0% of Al_2O_3, 11.5 to 15% of CeO_2, 6.2 to 7.4% of Tb_4O_7

and 5.5 to 9.5% of MgO. The blue phosphors consist of 61.2 to 71.0% of Al_2O_3, 13.8 to 21.0% of BaO and 4.5 to 5.3% of MgO. This yields a number of oxides (table 25) for recycling the constituent elements (table 26).

Table 25. Rare-earth oxides yielded by
aluminate-type tri-colour phosphors

Oxide	Content (%)
Y_2O_3	46.9–51.2
Al_2O_3	29.9–35.9
CeO_2	4.1–5.3
Eu_2O_3	3.9–4.4
MgO	2.7–4.0
Tb_4O_7	2.2–2.6
BaO	2.1–3.2

Table 26. Elemental composition of waste tri-colour phosphors

Element	Content (wt%)
silicon	17.89
calcium	12.44
barium	5.79
phosphorus	5.56
sodium	4.94
aluminium	4.32
yttrium	2.88
magnesium	1.02
strontium	0.73
potassium	0.47
fluorine	0.38
iron	0.38

lanthanum	0.31
manganese	0.30
cerium	0.30
chlorine	0.22
lead	0.21
antimony	0.19
europium	0.14
terbium	0.12
gadolinium	0.09

De-silication, decomposition and acidolysis have been used[99] to recover rare earths from scrap trichromatic phosphors which contained glass. Some **88%** of the glass can be removed by dry-sieving through a 0.05mm mesh and leaching using 5mol/l NaOH solution at 90C for 4h and a 5:1 liquid/solid ratio. Blue and green phosphors are decomposed by alkaline fusion (600C, 2h) and yttrium-, europium-, cerium- and terbium-rich solutions are obtained by 2-step acidolysis. The total leaching-rate of rare earths attains 94%, with the individual rates for yttrium, europium, cerium and terbium being 96, 99, 81 and 92%, respectively.

The phosphor dust which is obtained by crushing and sieving tubular lights contains some 34% of rare earths, such as yttrium, europium, cerium and terbium, as $Y_{1.90}Eu_{0.10}O_3$ and $Al_{11}Ce_{0.67}MgO_{19}Tb_{0.33}$ phases[100]. The recovery of yttrium and europium at rates greater than 95% is possible via leaching, followed by the recovery of 40% of cerium and more than 95% of terbium from leach residue by using microwave exposure followed by leaching. The leach residue is microwaved with NaOH in order to dissociate cerium- and terbium-bearing phases. It is possible to recover 52g of mixed oxides of yttrium and europium and 7g of mixed oxides of terbium, cerium, europium and yttrium - in purities of better than 99% - from 100 tubular lights.

Acid-leaching remains the most important chemical process for extracting rare earths from the waste tri-colour phosphors, and its efficiency increases with increasing temperature, sulfuric-acid concentration and agitation. The leaching efficiency of yttrium, europium, cerium and terbium is of the order of 80.4, 82.2, 81.4 and 80.0%, respectively, when performed at 100C using 2M H_2SO_4 over 8h.

Materials Research Forum LLC
https://doi.org/10.21741/9781644901793

The alternative alkali-fusion method involves the thermal decomposition of insoluble substances. The phosphors are first fired with an alkali under conditions which are sufficient to decompose them into a mixture of oxides. A residue which contains the rare-earth oxides is then extracted from the mixture and is treated so as to obtain a solution which contains rare earths in salt form. They are then separated from the solution. With a mass ratio of 6:1, a temperature of 900C and a reaction time of 2h, the leaching efficiency of yttrium, europium, cerium and terbium can approach 100%. The alkali-fusion method can effectively destroy the spinel structure of waste powders.

The rare-earth ions in solution can be turned into an insoluble precipitate by adding reagents such as oxalic acid. The addition of excess oxalic acid to rare-earth nitrate or chloride solutions can produce rare-earth oxalate precipitates:

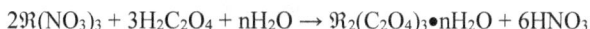

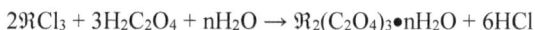

$$2\Re(NO_3)_3 + 3H_2C_2O_4 + nH_2O \rightarrow \Re_2(C_2O_4)_3 \bullet nH_2O + 6HNO_3$$

$$\Re_2(SO_4)_3 + 3H_2C_2O_4 + nH_2O \rightarrow \Re_2(C_2O_4)_3 \bullet nH_2O + 3H_2SO_4$$

$$2\Re Cl_3 + 3H_2C_2O_4 + nH_2O \rightarrow \Re_2(C_2O_4)_3 \bullet nH_2O + 6HCl$$

The solubility of a rare-earth oxalate can increase when sulfuric, nitric or hydrochloric acids are present in solution, and this increases with increasing acidity. For a given concentration of acid, the rare-earth oxalate solubility is highest in hydrochloric acid, next-highest in nitric acid and lowest in sulfuric acid. For a given acidity, the solubility of a rare-earth oxalate decreases with increasing atomic number (table 27).

Table 27. Solubility of rare-earth oxalates in water at 25C

Oxalate	Solubility (g/l)
$Sc_2(C_2O_4)_3 \bullet 6H_2O$	7.40
$Yb_2(C_2O_4)_3 \bullet 6H_2O$	3.34
$Sc_2(C_2O_4)_3$	3.11
$Y_2(C_2O_4)_3 \bullet 6H_2O$	1.00
$Nd_2(C_2O_4)_3 \bullet 6H_2O$	0.74
$Pr_2(C_2O_4)_3 \bullet 10H_2O$	0.74

$Sm_2(C_2O_4)_3 \bullet 6H_2O$	0.69
$La_2(C_2O_4)_3 \bullet 10H_2O$	0.62
$Gd_2(C_2O_4)_3 \bullet 6H_2O$	0.55
$Ce_2(C_2O_4)_3 \bullet 10H_2O$	0.41

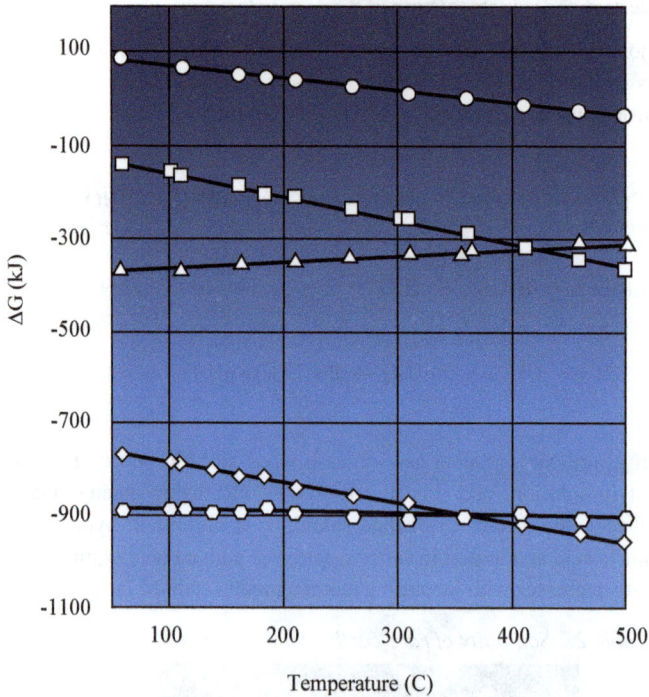

Figure 4. Free energy changes of main reactions involved in the selective chlorination of rare earths in NdFeB magnets. Circles: $NH_4Cl \rightarrow NH_3 + HCl$, squares: $Fe + 3NH_4Cl + 0.75O_2 \rightarrow FeCl_3 + 3NH_3 + 1.5H_2O$, triangles: $Fe + 0.75O_2 \rightarrow 0.5Fe_2O_3$, diamonds: $Nd + 3NH_4Cl + 0.75O_2 \rightarrow NdCl_3 + 3NH_3 + 1.5H_2O$, hexagons: $Nd + NH_4Cl + 0.75O_2 \rightarrow NdOCl + NH_3 + 0.5H_2O$

The method of chlorination-roasting assisted by alkaline fusion has been used[101] to extract rare earths from waste phosphors by using molten $AlCl_3$-KCl for the chlorination roasting and Na_2CO_3 for the alkaline fusion. The phosphors were first subjected to chlorination roasting in order to extract yttrium and europium. The products were subjected to secondary chlorination following alkaline fusion in order to extract cerium and terbium. The elements, in oxide form, could be selectively chlorinated to their chlorides while impurities such as silicon, aluminium and phosphorus could not be converted to chlorides by roasting with $AlCl_3$. The yttrium and europium in red phosphors were easily chlorinated by molten $AlCl_3$-KCl, while the green and blue phosphors were hardly attacked by the molten chlorides. The extraction efficiencies of yttrium and europium were therefore greater than 96%, while those of cerium and terbium were lower than 30% for 1-step chlorination-roasting at 700C for 3h, given a chloride/phosphor mass-ratio of 4:1. The yttrium and europium in red phosphors were almost completely extracted, whereas green and blue phosphors plus $Ca_5(PO_4)_3F_{0.94}Cl_{0.1}$ and SiO_2, remained in the roasted product. The latter were then mixed with an equivalent mass of Na_2CO_3 and roasted (1000C, 3h). The rare earths in aluminate green and blue phosphors could be converted into their oxides by alkaline fusion. These products were in turn again processed by chlorination-roasting. The cerium and terbium in the waste phosphors could be easily converted to their chlorides and the extraction efficiencies were increased to 88.51% and 83.06%, respectively, during the secondary chlorination-roasting. In the case of 1-step chlorination, the overall extraction efficiency of yttrium, europium, cerium and terbium was 89.15% under optimum conditions. The extraction efficiency could be increased to better than 97% when the phosphors were treated by molten-salt chlorination followed by alkaline fusion and secondary chlorination The conventional oxidation-roasting and acid-leaching method has been changed[102] to a lower-temperature chlorination-roasting and water-leaching process by reducing the roasting temperature to 300C (figure 4). Ammonium chloride was used as a chloridizing agent. The optimum recovery conditions were a temperature of 300C and 3h, followed by water leaching. Rare-earth oxide of 99.2% purity was produced from the leachate. Another chlorination process using zinc chloride has been investigated[103] for the selective recovery of rare earths. A mixture of Nd-Fe-B magnet powder and $ZnCl_2$ was placed in a gas-tight quartz tube and pre-heated to 1000K over a period of 1.5 to 5h. Under certain conditions, the chlorination efficiencies of neodymium, dysprosium and praseodymium were 96.5%, 57.2% and 97.6%, respectively. Iron and neodymium chloride were generated.

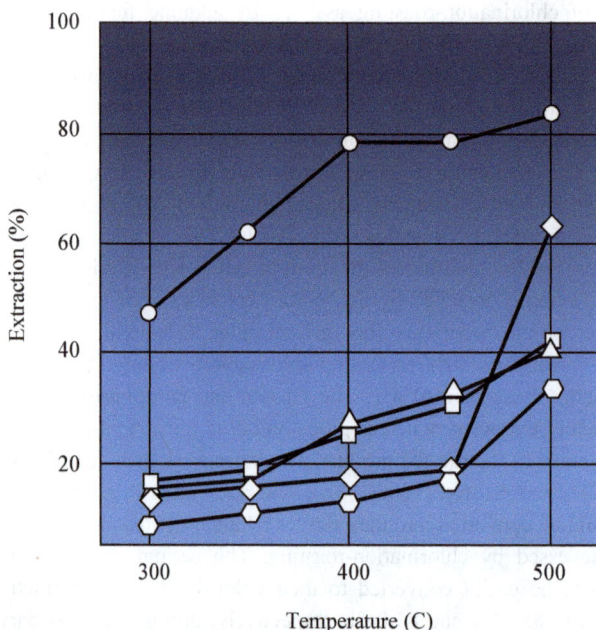

Figure 5 Effect of roasting temperature on extraction of metals during first roasting stage. Circles: $\Re$, squares: Co, triangles: Cu, diamonds: Fe, hexagons: Al

The use of a 2-stage roasting process (figures 5 and 6) can greatly reduce the amount of ammonium sulfate that is involved and improve the separation efficiencies of rare earths[104]. During first low-temperature roasting, almost 80% of the rare earths can be transformed into $\Re_2(SO_4)_3$ or $NH_4\Re(SO_4)_2$ within 1h at 400C. Iron and other impurities are simultaneously converted into insoluble metal ammonium sulfates. These products are then further roasted at 750C for 2h, leading to the extraction of up to 96% of the rare earths. Extraction of related impurities such as iron, aluminium, copper and cobalt amounted only to 0.008%, 0.27%, 1.64% and 3.48%, respectively. The decomposition of $NH_4\Re(SO_4)_2$, and the reaction of $Fe_2(SO_4)_3$ and $\Re_2O_3$, together improved rare-earth extraction during the second stage.

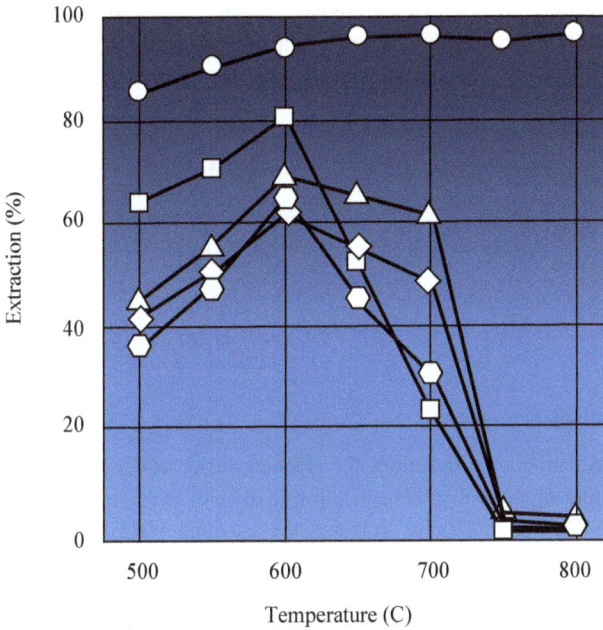

Figure 6. Effect of roasting temperature on extraction of metals during second roasting stage. Circles: ℜ, squares: Fe, triangles: Co, diamonds: Cu, hexagons: Al

A non-polluting closed-loop process has been developed[105] for the treatment of rare-earth concentrates. By means of an oxidative-roasting hydrochloric-acid leaching step, more than 98% of fluorine and phosphorus can be retained in the leaching residue. The latter is then converted into rare-earth hydroxides, sodium fluoride and sodium phosphate in sub-molten sodium hydroxide. Most of the concentrated hydroxide can be re-used after filtration. The washed filter residue is dissolved in hydrochloric acid so as to produce a rare-earth chloride solution. Sodium fluoride and sodium phosphate can be recovered from the washing solution. The overall process thus cycles concentrated hydroxide and water while producing rare-earth chloride solutions, sodium fluoride and sodium phosphate. The rare-earth concentration in the chloride solution reached 280g/l, while 96.40% of the fluorine was converted into sodium fluoride and 99.82% of the phosphorus was converted into sodium phosphate.

Leaching can also be used to extract rare earths, especially neodymium and dysprosium, from scrap magnets. Neodymium can be leached by using various HCl, H_2SO_4, HNO_3 or NaOH. The components of the magnets dissolve according to:

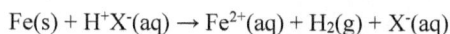

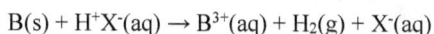

$$Nd(s) + H^+X^-(aq) \rightarrow Nd^{3+}(aq) + H_2(g) + X^-(aq)$$

$$Fe(s) + H^+X^-(aq) \rightarrow Fe^{2+}(aq) + H_2(g) + X^-(aq)$$

$$B(s) + H^+X^-(aq) \rightarrow B^{3+}(aq) + H_2(g) + X^-(aq)$$

The leaching efficiency of NaOH tends to be lower than that of the acids.

An increase in temperature decreases the leaching efficiency and the linear dependence upon temperature suggests that the leaching-rate of acids is controlled by the kinetics. An increase in the solid/liquid ratio decreases the leaching yield because the increase in density leads to a lower availability of reagent per unit weight of waste. On the other hand, an increase in acid concentration increases the dissolution of rare earths from magnet scraps. A 3M acid concentration, a density of 2%(w/v) and a temperature of 27C are the optimum conditions for leaching out more than 95% of the neodymium from Nd-Fe-B waste within 0.25h.

The leaching efficiency of $(NH_4)_2SO_4$ and H_3PO_4 is only 40% for dysprosium and less than 5% for neodymium. The addition of H_2O_2 can enhance the leaching efficiency and HCl or HNO_3 could, in the presence of H_2O_2, could leach out more than 95% of the rare earths. Nitric acid is the better leachant because, paradoxically, of its inferior ability to leach out iron. This is because the presence of iron in the leachate impairs the recovery of rare earths. It is possible to recover neodymium selectively from Nd-Fe-B magnet scrap leachate by altering the pH level. The pH of the leachate is generally between 0.13 and 0.02, but can be adjusted by using sodium hydroxide. At a pH of 0.6, more than 95% of the neodymium can be recovered, as hydroxide precipitates, from sulfuric acid leached Nd-Fe-B magnet waste. The selective recovery of neodymium from HCl-leached Nd-Fe-B magnet waste is not so successful.

Various hydrometallurgical processes have been proposed for the leaching of rare earths from scrapped nickel metal hydride batteries, with the usual mineral acids generally being chosen. Again in some cases, H_2O_2 is added to the acids. By using 1M hydrochloric, sulfuric and nitric acids at a pH level of 1.0 and a temperature of 30C the

Materials Research Forum LLC
https://doi.org/10.21741/9781644901793

anodic parts of the batteries could be completely dissolved within 6h. The use of 8M HCl is optimum for the leaching of rare earths from a cathode and anode mixture. One two-step process for the efficient leaching of lanthanum and cerium from battery scraps involves, in the first step, leaching of the electrode material with 3M H_2SO_4 at 80 to 85C for 3h. In the second step, leaching is carried out using 1M H_2SO_4 at 20C for 1h. In another two-step process for cobalt and rare earths, the first step is to wash in water at 95C for 1h and then roast at high temperatures for 4h. The second step is to leach the roasted product with H_2SO_4 solution at 90C for 6h. This leads to 98% recovery of cobalt and rare earths. A further two-step process for the leaching of lanthanum, cerium, neodymium, praseodymium and samarium from the batteries involves first baking with 2ml of H_2SO_4 at 300C for 1.5h. This pre-treatment transforms nickel, zinc and rare earths into their sulfate form. Following baking the first step of leaching is carried out using water at 75C. This leaches out 91% of the nickel, 94% of the zinc and 91% of the rare earths. Some 20 to 30% of cobalt, iron and manganese is also leached out. The second step is carried out using $NaHSO_3$ in H_2SO_4 at 95C, and leaches out residual cobalt and manganese. The precipitation characteristics of rare earths with sulfate have been investigated[106] and compared with those of precipitants such as phosphate, oxalate and fluoride/carbonate. The precipitates could be anhydrous sulfate, octa-hydrated sulfate and sodium double salt; which was in turn compared with anionic double-salt precipitation by fluoride-carbonates. It was found that anions such as Cl^-, NO_3^- and SO_4^{2-} play an important role in precipitation, due largely to complexation with the dissolved rare earths. The general precipitation effectiveness was in the order: sodium double salt, hydrated sulfate, anhydrous sulfate. The synergistic role of double-salt precipitation, cationic or anionic, was often as effective as that of oxalates and phosphates, even at low pH levels.

Oxalate precipitation can be used to obtain rare earths from leachates: for example, the precipitation of yttrium from spent fluorescent lamp waste leachate. Nitric acid produced magnet leachate contains 28g/l of neodymium, 0.7 to 0.8g/l of dysprosium, 4.0 to 6.6g/l of iron and 0.4 to 0.5g/l of boron. The iron is first removed as $Fe(OH)_3$ by adding sodium hydroxide at a pH of 2.0 to 3.0, although 20 to 25% of any rare earths present are co-precipitated with, or trapped in, the iron precipitate. Oxalic acid is then used to precipitate rare earths as oxalates from the iron-depleted leachate. More than 70% of the neodymium can be recovered as $Nd_2(C_2O_4)_3 \bullet 10H_2O$ by using 1.1M oxalic acid.

Simple water-soluble aminobis(phosphonate) ligands, $XN[CH_2P(O)(OH)_2]_2$, where X was CH_2CH_3, $(CH_2)_2CH_3$, $(CH_2)_3CH_3$, $(CH_2)_4CH_3$, $(CH_2)_5CH_3$ or $CH_2CH(C_2H_5)(CH_2)_3CH_3)$, have been proposed[107] as precipitating agents. The latter 3 agents were found to separate thorium, uranium and scandium within 0.25h, with the separation factors ranging from 100 to 2000 in acidic aqueous solutions. All of the agents improved the separation factors for adjacent lanthanoids, unlike traditional oxalate precipitation agents. The precipitated metals could be recovered from the agents by using 3M HNO_3, with no apparent ligand decomposition.

Mini-hydrocyclones have been used[108] to separate fine rare-earth particles from suspensions at flow-rates of 1200 to 1600l/h, split ratios of 20 to 60% and concentrations of 0.6 to 1.0wt%. The optimum parameters required to ensure a total separation efficiency of 92.5% were a flow-rate of 1406l/h, a split ratio of 20% and a feed concentration of 1wt%.

Solvent extraction

The solvent extraction technique involves the use of two liquid phases which are completely immiscible. Solvent extraction is the most commonly used technique for the purification of rare earths because it is able to handle large volumes of material and maintain a high product purity. The scalability of the membrane solvent extraction process for the recovery of rare earths from scrap permanent magnets has been demonstrated by processing larger quantities of magnet scrap with the use of a membrane area of more than $1m^2$. The membrane solvent extraction process could recover high-purity rare earths, in oxide form, from a wide range of end-of-life magnet waste material[109]. Rare earths with a purity of more than 99.5wt%, with a recovery efficiency of more than 95% and an extraction rate as high as $9.3g/m^2h$ were recovered from solutions which contained rare-earth concentrations of up to 46000mg/l. The extraction rate depended markedly upon the initial rare-earth concentration in the solution. An empirical equation has been used[110] to correct the non-ideality of the organic phase and calculated extraction equilibrium constants for trivalent rare earths, lanthanum, cerium, praseodymium, neodymium, samarium, europium, terbium, dysprosium, yttrium, which are especially important in recycling. Calculations which were based upon use of 2-ethylhexylphosphonic acid mono-2-ethylhexyl ester in a non-polar diluent and a nitrate medium were used to investigate the equilibria at a constant ionic strength of 1mol/l at 298K. The 2-ethylhexylphosphonic acid mono-2-ethylhexyl ester concentrations ranged from 0.04 to 0.33mol/l, for a range of pH values. Apparent extraction equilibrium constants were obtained which accurately reproduced the distribution ratios of rare earths. The use of higher 2-ethylhexylphosphonic acid mono-2-ethylhexyl ester concentrations revealed that the empirical equation was effective up to a concentration of 0.5mol/l; a considerably higher than the accepted upper bound. Distribution ratios of two rare earths were calculated from the extraction equilibrium constants. Differences in the separation factors depended upon the measurement conditions and were larger when the difference in the atomic numbers of the rare earths was large. The distribution coefficients of Y^{3+} and Eu^{3+} ions between an organic phase, di-(2-ethylhexyl) phosphoric acid, and aqueous nitric acid solution were measured[111] as a function of the nitric acid concentration at about 298K. The organic phase was dissolved in n-nonane or n-dodecane. The distribution coefficients were inversely proportional to the aqueous acid concentration, and depended upon the di-(2-ethylhexyl) phosphoric acid concentration and the type of diluent.

The separation of neodymium and dysprosium is possible by using ionic liquids for solvent extraction. Rare earths can be recovered from waste Nd-Fe-B magnet leachate by using bis(2-ethylhexyl) phosphoric acid. The Nd-Fe-B magnets are first sulfated, roasted and leached with water so as to dissolve the rare earths and give a leachate which contains 9.1mM of neodymium, 2.7mM of dysprosium, 3.2mM of praseodymium, 0.69mM of gadolinium, 0.17mM of cobalt and 0.55mM of boron. At this stage, the iron concentration is undetectable. bis(2-ethylhexyl) phosphoric acid is then used as an organic extractant in concentrations of 0.3, 0.6, 0.9 or 1.2M. All of the rare earths are separated as a group, with aliphatic diluents offering a greater extraction efficiency than do polar diluents. The use of 0.3M bis(2-ethylhexyl) phosphoric acid in hexane is the best choice for ensuring the maximum extraction and separation of neodymium, dysprosium and praseodymium. Ionic liquids have the advantage of being environmentally friendly solvents which possess an extremely low volatility and combustibility and offer so-called tuneability[112]. The selective recovery of metals by ionic liquids depends upon factors such as the metal and water contents of the feed-stock. Rare earths can be separated from other metals in fluorescent-lamp phosphor leachate by using N,N-dioctyldiglycolamic acid as an extractant and [C4mim][NTf2] as a solvent. Undiluted trihexyl(tetradecyl)phosphonium chloride can be used for the separation of iron, cobalt, copper, manganese and zinc from neodymium and samarium. Dialkylphosphate ionic liquid can be used to separate neodymium from nitric acid magnet leachate, and neodymium and dysprosium can be selectively recovered from waste magnets by combining the ionic liquid, trihexyl(tetradecyl)phosphonium nitrate, and a selective complexing agent: ethylenediaminetetraacetic acid. By using trihexyl(tetradecyl)phosphonium chloride or tricaprylmethylammonium chloride ionic liquids, it is possible to separate cobalt, manganese, iron and zinc from rare earths. Trihexyl(tetradecyl)phosphonium nitrate gives a good separation of cobalt and samarium and of nickel and lanthanum. A recent innovation is the use of bifunctional ionic liquid extractants. Modification of well-known extractants such as di(2-ethylhexyl)phosphoric acid and 2-ethyl(hexyl) phosphonic acid mono-2-ethylhexyl ester can produce bifunctional ionic liquids that can separate well the early and late rare earths.

A comparison has been made[113] of the carbon footprints of yttrium and europium recovery from phosphors when using acid or solvent extraction. The same amounts of phosphor for specific yttrium and europium recovery concentrations, and the same amounts of phosphor for extraction were assumed. In the case of acid extraction, H_2SO_4 or HCl were used at 60 or 90C. In the case of solvent extraction, acid leaching was followed by ionic liquid extraction. The carbon footprints of the acid and solvent extraction methods were estimated to be 10.1 and $10.6kgCO_2$, respectively. Solvent extraction offered a much higher efficiency, although acid extraction involved a lower carbon footprint.

There is interest in developing novel ionic liquids as possible substitutes for the usual organic solvents used for extracting rare earths. Carboxylic acid functionalized phosphonium-based ionic liquids, (4-carboxyl)butyl-trioctyl-phosphonium chloride and (4-carboxyl)butyl-trioctyl-phosphonium nitrate, have been synthesized[114]. Samples were tested as undiluted hydrophobic acidic extractants for rare-earth ions, using a maximum loading of 3mol/mol of Nd^{III} in aqueous solution with a notable stripping ability. The results demonstrated excellent selectivity of Sc^{III} from among mixtures of 6 rare-earth ions. There was also a remarkable separation of rare-earth ions and first-row transition-metal ions: La/Ni, Sm/Co. The extraction mechanism involved proton exchange in the ionic liquid phase.

The ionic liquid, pyridine hydrochloride, is a suitable non-aqueous solvent for metal oxides such as those of rare earths but its use is limited by its miscibility with the aqueous phase. Molten pyridine hydrochloride at 165C was used[115] to dissolve scrap Nd–Fe–B permanent magnets in order to recover neodymium and dysprosium. The powdered scrap completely dissolved within 10min, with a lixiviant/solid ratio of 10. Non-aqueous solvent extraction was then carried out 165C by using PC-88A molecular extractant or Cyphos IL101 ionic liquid. The high temperature which was used had the effect of lowering the viscosity of the solvents so that they could be used in undiluted form. The high temperature also affected the equilibrium constants and therefore the distribution of metals between the two phases. In a first stage, 30vol% of PC-88A in p-cymene was used to extract Dy^{III}. In a second stage, 100vol% of PC-88A was used to extract most of the Nd^{III}. A mixture of Cyphos IL101 and p-cymene in a 70:30 ratio could efficiently extract $Fe^{II,III}$ from leachates.

An alkali fusion, acid leaching and liquid–liquid extraction process has been proposed[116] for the separation of lutetium from $(Lu,Y)_2SiO_5$ scrap by using diglycolamide-based ionic liquids. The process could extract rare earths under low-acidity conditions during hydrochloric acid leaching. Two ionic liquids were used to extract Lu^{III} and Y^{III} from a chloride system. Stripping of the Lu^{III} from the organic phase could be achieved by using 0.01mol/l HCl solution.

Materials Research Forum LLC

https://doi.org/10.21741/9781644901793

Consumer-goods sources

Batteries

Mixed scrap alkaline, Li-ion, nickel metal-hydride and Ni-Cd batteries have been heat-treated[117] at various temperatures, concentrated by screening and subjected to sulphuric-acid oxidation leaching with added H_2O_2 and $Na_2S_2O_8$. These additions increased the leaching efficiency of nickel up to 98%, as compared with sulfuric acid leaching. Lanthanum was also selectively and completely precipitated as $La(SO_4) \bullet 2Na \bullet H_2O$.

Nickel metal hydride battery waste can in fact be use[118]d as a reductant for lithium-ion battery waste, thereby synergistically improving the extraction of metals from the 2 types of waste. The main benefit of the process was the reduced consumption of leaching chemicals. Crystallization of sodium sulfate was found to be the most environmentally feasible option, as it permitted the use of sodium as a precipitant for rare-earth recovery.

The optimum conditions for the leaching of rare earths from spent batteries were found[119] to be 2M H_2SO_4, 348K and 2h for a pulp density of 100g/l. Under these conditions, leaching of 98.1% of neodymium, 98.4% of samarium, 95.5% of praseodymium and 89.4% of cerium was possible. More than 90% of the base metals (nickel, cobalt, manganese, zinc) were also leached out. Activation energies of 7.6, 6.3, 11.3 and 13.5kJ/mol were deduced for the leaching of neodymium, samarium, praseodymium and cerium, respectively, at 305 to 348K. Mixed rare earths were precipitated from leachant at a pH level of about 1.8.

In a nickel metal-hydride battery, metallic components account for more than 60% of its weight[120]. The contents of nickel, iron, cobalt and rare earths in a single battery are 17.9, 15.4, 4.41 and 17.3%, respectively. Some 1.88g of rare earths, including cerium, lanthanum and yttrium, can be obtained from one battery. Metal extraction from scrap nickel metal-hydride batteries can be achieved[121] by using 1M H_2SO_4 at a pulp density of 25g/l and 90C. More than 99% of the rare earths is precipitated out from leachant at a pH level of 1.8 by using 10M NaOH, and is isolated by calcination at 600C. Undesired metals such as manganese, aluminium, zinc and iron are scrubbed from leachant by using 0.7M di(2-ethylhexyl) phosphoric acid at a pH level of 2.30.

An early hydrometallurgical method[122] for the recovery of rare earths and transition metals from the negative electrodes of scrap Ni-metal-hydride cell-phone batteries involved chemical precipitation, at a pH level of 1.5, of sodium cerium sulphate, $NaCe(SO_4)_2 \bullet H_2O$, and lanthanum sulphate, $La_2(SO_4)_3 \bullet H_2O$. Iron was meanwhile recovered as $Fe(OH)_3$ and FeO and manganese was recovered as Mn_3O_4; with nickel and cobalt oxides being recycled as cathodes for Li-ion batteries. In another process[123], electrodes from the scrap batteries were leached using H_2SO_4; with ozone as an oxidant.

Materials Research Forum LLC

https://doi.org/10.21741/9781644901793

Recoveries of 96% were possible for lanthanum, cerium and neodymium at room temperature while nickel and cobalt were electrochemically separated from the leach solution and rare earths were precipitated. In the particular case of lanthanum, its selective extraction from Ni-metal-hydride battery leachate was achieved[124] by using aqueous two-phase systems. The processes were evaluated with regard to the effect of the concentration of various extractants, the pH level and electrolyte (Li_2SO_4, Na_2SO_4, $MgSO_4$, $Na_2C_4H_4O_6$, $Na_3C_6H_5O_7$) type. The best conditions for extraction were found to be the use of a polymer (PEO1500) plus Li_2SO_4 at a pH level of 6.00; with 1,10-phenanthroline as the extractant. This provided an extraction efficiency of 74.1%. Following 3 successive extraction steps, the separation factors for the separation of lanthanum from other metals were La/Ce = 180, La/Pr = 184 and La/Nd = 185. Another process[125] uses super-critical fluid CO_2 solvent extraction to recover the 30wt% of rare earths (lanthanum, cerium, praseodymium, neodymium) which are present in the anodes of Ni-metal-hydride batteries. This process recovers some 90% of the rare earths, operates at low temperatures and does not produce hazardous wastes. The mechanism of super-critical fluid extraction of rare earths is supposed[126] to involve a trivalent state which bonds with three tri-n-butyl phosphate molecules and three nitrates in the extracted complexes. Crushed battery scrap can be characterized[127] in terms of element distribution per particle size and density. Good separation of iron and plastics can be obtained by using a 1mm sieve, and an acid consumption of 14mol H^+ ions per 1kg of scrap is sufficient to achieve a desired final pH level of less than 1. Leachant which was rich in nickel (46g/l), lanthanum (9g/l), cerium (7.5g/l), praseodymium (1.4g/l), samarium (0.29g/l) and yttrium (0.17g/l) was obtained, and rare-earth precipitation was investigated as a function of dilute (0.01 to 0.5M) Na_2SO_4 solution content at 50C. The best precipitation efficiency was found for a sodium to rare-earth ratio of 9.1. This resulted in a better than 99% precipitation efficiency for rare earths. The separation of rare earths and transition metals is a critical step in the recycling of nickel-metal hydride batteries, and the ionic liquid, [trihexyl(tetradecyl)phosphonium]$_2$[2,2'-(1,2-phenylenebis(oxy))dioctanoate] ([P6,6,6,14]$_2$[OPBOA]), has been newly synthesized[128] for this purpose. The separation factors of Nd/Co and Nd/Ni are of the order of 5.2. x 10^3 and 5.4. x 10^3, respectively. The extracted nickel and cobalt can be effectively stripped by using 0.5mol/l NaCl solution without any loss of the rare earth. The neodymium in [P6,6,6,14]$_2$[OPBOA] could be stripped by using 0.014mol/l HCl, 0.016mol/l $Na_2C_2O_4$ or 0.022mol/l Na_2CO_3. The [P6,6,6,14]$_2$[OPBOA] could also be recycled without regeneration when $Na_2C_2O_4$ and Na_2CO_3 were used as stripping agents. Larger precipitates were obtained when using dilute Na_2CO_3 solution. In the case of 500ml of feed solution, the recovery efficiency and purity of rare earths could attain more than 96.4

and 99.8wt% when using NaCl and Na_2CO_3, respectively. Rare earths in oxide form could then be easily obtained by calcining.

Lanthanum, cerium, neodymium and praseodymium have been isolated in oxide form from Ni-metal hydride by using an oxidation-reduction process[129]. The anode part, with a composition of 54wt%Ni, 23.7wt%La, 6.7wt%Ce, 5.4wt%Co, 3.6wt%Nd and 3.4wt%Mn, was treated by using an oxidation process in air at 1000C for 1h. This was followed by reduction at 1550C for 1.5h by using iron as a reducing agent. This resulted in the separation of iron-based and rare-earth oxide phases. The latter were rich in lanthanum, cerium, neodymium and praseodymium.

The leaching of nickel and rare earths in spent nickel metal hydride battery powders was investigated[130] in HCl and H_2SO_4 at 25 to 60C for pH levels of 3 to 5.5. Anomalous particles were observed which exhibited a core–shell structure that was related to anode active-mass aging. The selective dissolution of rare earths, compared with that of nickel, at a pH level of 3 was attributed to the kinetic inhibition of nickel metal dissolution and to the above core–shell structure of aged mischmetal particles. The use of H_2SO_4 led to the co-precipitation of lanthanide–alkali double-sulfates and nickel salts. A better understanding of the precipitation process was obtained from small-scale experiments on 21 samples of industrial leach solutions which contained 50g/l of nickel and 17g/l of rare earths[131]. Thermodynamic modelling predicted the influence of temperatures of 25C to 60C when the sodium/rare-earth molar ratio was between 0.8 and 3.2. Very selective precipitation was observed at 60C when the Na/$\mathfrak{R}$ molar ratio was 4:1.

Mixed alkaline rare-earth double-sulfate precipitate which resulted from the sulphuric-acid leaching of nickel-metal hydride battery waste was used[132] to study the process via which double sulfates are transformed into hydroxides, with the simultaneous *in situ* conversion of Ce^{III} into Ce^{IV} by air. Air flow-rates ranging from 0 to 1l/h, temperatures ranging from 30 to 60C, liquid/solid ratios of 12.5 to 100g/l and times of 1 to 4h were used to study oxidation and double-sulfate conversion. The best degree of oxidation was 93%, together with almost complete dissociation of the double-sulfate matrix; 52767ppmNa being reduced to 48ppmNa. The selective dissolution of rare earths in HNO_3 led to an end-product of concentrated impure $Ce(OH)_4$.

The precipitation of rare earths from sulfate media is often attributed to an increase in the pH level of the leaching solution, but it also depends upon the Na^+ and SO_4^{2-} concentrations as well as the pH. In a study[133] of the 2-stage leaching of crushed Ni-M-H waste, the first stage involved leaching with 2M H_2SO_4 at 30C. The second stage involved H_2O leaching at 25C. A higher-than-stoichiometric quantity of sodium salts was then used during precipitation. The increase in precipitation and reagent concentrations led to an improved efficiency of double-sulfate precipitation. The best rare-earth precipitation efficiencies of 98 to 99% were attained by increasing the concentrations of H_2SO_4 and Na_2SO_4 by 1.59M and 0.35M, respectively. This resulted in a 21.8 times Na and 58.3 times SO_4 change in the stoichiometric ratio with respect to the rare earths.

On the basis of the differing solubilities of metal salts, a step-wise leaching process was proposed[134] for the recovery of rare earths from spent battery materials. Over 99% of the rare earths, cobalt, nickel and manganese were leached out, with the leaching kinetics obeying a shrinking-core model and interdiffusion being the rate-determining step. The rare earths were recovered in the form of high-purity sulfuric acid complex salts, while nickel, cobalt and manganese were used to synthesize $LiNi_{0.6}Co_{0.2}Mn_{0.2}O_2$ cathodes.

A high-temperature process for recycling scrap nickel metal-hydride batteries was developed[135] in which the positive and negative electrodes, plus polymer separator, were heated to between 600 and 800C in order to remove organic components and isolate the nickel-based negative electrode. The heat-treated materials thus consisted mainly of nickel-, rare-earth- and cobalt- oxides. The rare-earth oxides were recovered by using a high-temperature treatment in which slags consisting mainly of SiO_2 and CaO were used as rare-earth oxide absorbents. Following this treatment, over 98% of the nickel and cobalt oxides were reduced to metal while nearly all of the rare-earth oxides remained in the molten slag; where they selectively precipitated in the form of solid SiO_2-CaO-Re_2O_3, with the slag matrix being Re_2O_3-deficient to the extent of less 5wt%.

A study[136] of the recycling of scrap nickel metal-hydride batteries showed that an increase in temperature, hydrochloric acid concentration and leaching-time increased the leaching-rate of rare earths. A maximum rare-earth recovery of 95.16% was possible under the optimum leaching conditions of 70C, solid/liquid ratio of 1:10, 20%HCl concentration, 74μm particle size and 100min leaching time. The experimental data could be explained by invoking a reaction-controlled process with an activation energy of 43.98kJ/mol. Following leaching and filtration, rare-earth oxalates could be obtained by adding saturated oxalic solution to the filtrate. After removing any impurities by adding ammonia, filtering, washing with dilute HCl and calcining at 810C, the final product consisted of 99%-pure rare-earth oxides.

Catalysts

Scrap catalysts can contain more than 2% of rare earths, especially lanthanum and cerium. A solution of rare-earth chlorides, containing both rare earths and impurities is obtained by leaching with hydrochloric acid. Lanthanum and cerium have been extracted[137] from a hydrochloric acid system by using 2-ethylhexyl phosphonic acid mono 2-ethylhexyl. The process involved a 2-ethylhexyl phosphonic acid mono 2-ethylhexyl volume concentration of 60vol%, a leaching-solution pH of 2.5, an organic-phase/water-phase volume ratio of 2:1 and an equilibrium extraction time of 0.5h. The solution of rare-earth chloride from the loaded organic phase could be extracted by using hydrochloric acid, with the back-extract concentration of hydrochloric acid being 2.0mol/l and the equilibrium time for that extraction being 1h.

Investigation of the hydrometallurgical recycling of cerium from the secondary residue of scrap autocatalysts has shown[138] that cerium dissolution is improved by up to 96% when hydrofluoric acid is added to 2.0mol/l sulfuric acid solution. An activation energy of 31.8kJ/mol suggested that a diffusion-controlled mechanism governed the cerium leaching. Leachant which contained 4.2g/l of cerium was later equilibrated with Cyanex 923 in order to dissolve relevant components into an organic phase. The affinity of Cyanex 923 for cerium was spontaneous and was governed by an energy of -6.58kJ/mol at 298K. It exhibited outer-sphere coordination with regard to the exothermic process (-21.42kJ/mol). The cerium was removed, with better than 98% efficiency, from the organic phase by using a mixture of 1.0mol/l of H_2SO_4 and 0.5mol/l of H_2O_2. High-purity $Ce_2(C_2O_4)_3 \bullet 10H_2O$ was eventually precipitated by adding oxalic acid at an optimum concentration of $Ce^{3+}:H_2C_2O_4$ of 1:5.

Microwave-assisted leaching has been used[139] to separate platinum-group metals and light rare-earth elements from end-of-life automobile ceramic catalyst materials in 6M HCl at 150C. Hydrogen peroxide solution (10vol%) was added in some cases. Gas which was generated in the pressure-tight reactor, and speciation in the catalysts, affected the leachability of the platinum-group metals and the light rare earths. The formation of chlorine in the headspace furnished the 6M HCl system with a suitable oxidizing environment for leaching-out the platinum-group metals as soluble chloro-complexes. A spent catalyst which contained mainly oxidized platinum-group metals (93.9%Pd, 98%Pt, 70.7%Rh) was leached best by 6M HCl. The peroxide additions slightly decreased the platinum-group leaching efficiency, due to surface passivation. Spent catalysts which contained oxidizable species, such as Ce^{3+}, that gave rise to hydrogen evolution partially compensated the oxidation potential of the HCl system. In this case peroxide addition slightly improved the platinum-group leachability (91.8%Pd, 96%Pt, 89.9%Rh). Among the rare earths, cerium leaching was affected mainly by the passivation of Ce^{3+} due to oxidation. In the absence of peroxide and at low initial Ce^{3+} concentrations, the cerium was leached best (87 to 94%). The effect of peroxide was negligible in the case of lanthanum and neodymium and moderate in the case of yttrium. The leaching of the elements was impeded by their association with aluminium and zirconium oxides. A combined acid-leaching and oxalate-precipitation process has been used[140] to recover lanthanum from spent catalysts by using nitric acid rather than hydrochloric or sulfuric acid. The nitric acid was capable of completely leaching lanthanum. When combined with an oxalate precipitation step, high-purity (> 98wt%) lanthanum solid could be recovered. Electrokinetic remediation, combined with the leaching of spent catalysts, shows promise as an alternative method for the recycling of rare earths[141]. Use of sulfuric acid (1mol/l) and an applied electric field of 0.15V/m for 8h was most efficient with regard to energy and acid consumption per weight of lanthanum recovered.

Cathode-ray and liquid-crystal display screens

The main interest in recycling these items stems from the fact that the lead content of the glass is an environmental threat. So as well as being valuable in their own right, the rare earths extracted from the tubes can offset the overall cost of recycling. A process for simultaneously recycling rare earths and zinc from waste cathode ray tube phosphors involved first removing 95% of the glass and aluminium parts by simple screening[142]. A self-propagating high-temperature reaction and water-leaching process was then used to recycle zinc selectively. The rare earths were recycled by oxidative leaching and ionic-liquid extraction before being regenerated to form a new $Y_2O_3:Eu^{3+}$ phosphor. The recovery efficiencies of rare earths and zinc attained 99.5% and 99%, respectively. Any bivalent sulfur in ZnS and $Y_2O_3:Eu^{3+}$ was entirely converted to SO_4^{2-}, thus avoiding secondary pollution.

Yttrium was recovered from cathode ray tubes that had been manually dismantled, and the resultant powder was leached with HNO_3. The solution was then subjected to solvent extraction using di-(2-ethylhexyl) phosphoric acid, with n-heptane as a diluent. The HNO_3 was used again, and yttrium was precipitated by adding 4 times the stoichiometric amount of oxalic acid, leading to 68% purity yttrium[143]. Microwave-assisted leaching has been used[144] to recover yttrium and europium, with sulfuric acid being used as the leaching agent. A higher leaching efficiency was observed when the microwave power was increased from 200 to 600W and when the acid concentration was increased from 0.5 to 2mol/l. The leaching efficiencies of yttrium and europium were 78.07% and 100%, respectively, within 1h at a microwave power of 400W, using 2mol/l of H_2SO_4 and a 10g/l solid/liquid ratio. The thermal response and dissociation kinetics of the phosphors have been investigated[145], revealing that the relevant activation energy at 600 to 930C is

105.4kJ/mol. Recycling of the phosphors can yield 4.5g of yttrium, europium and lanthanum per 500kg of cathode ray tubes.

LCD screen wastes have been treated by using ultrasound-assisted leaching. The waste was first milled, and then sieved so as to pass through a 44μm mesh. The milled powder was subjected to ultrasound-assisted leaching for 1h in an aqueous medium with a pH of 6 at 25C. Magnetic separation was subsequently applied to the leach residue[146]. The waste was composed mainly of amorphous oxides of silicon, iron, indium, tin and rare earths; with the contents of gadolinium and praseodymium amounting to 93 and 24mg/kg, respectively. X-ray diffraction analysis of the magnetic fraction of the residue revealed the presence of amorphous phases together with crystalline metallic iron alloy. The formation of the latter was attributed to the effect of high-power ultrasonic solicitation during leaching. It was found that the magnetic residue represented 87wt% of the gadolinium and 85wt% of the praseodymium which was present in the original material.

Fluorescent lamps

There was little interest in the recycling of phosphors before 2009, but interest increased sharply during the following decade. This was largely because rare-earth phosphors, developed some 40 years earlier, became commercially favoured due to environmental legislation. It was initially concluded[147] that 2-liquid flotation, using organic phases, was an effective method for separating fluorescent powders by selecting suitable non-polar and polar solvents. During the first stage, the green phosphors were separated and, in the second stage, the blue powder was separated from the red powder in the presence of sodium 1-octane sulfonate (2×10^{-4}mol/l). The grade and recovery of each product was between 90 and 95% and the efficiency attained about 63%.

Metals are generally leached by using nitric acid, hydrochloric acid. sulphuric acid or ammonia in various processes. Ammonia is not suitable for the recovery of yttrium, while nitric acid produces toxic fumes. The best extraction of yttrium is obtained by using 20% 4N H_2SO_4 at 90C. The yttrium and calcium yields are some 85% and 5%, respectively. The acid concentration alone, and the interaction between acid and pulp density, have a markedly positive effect upon yttrium solubilization in both HCl and H_2SO_4. At least a stoichiometric amount of oxalic acid is required in order to recover yttrium efficiently, and 99% yttrium oxalate n-hydrate is then produced[148].

A solvent extraction method has been used[149] to recover yttrium from the leachant that arises from treating fluorescent lamp waste powder that has been dissolved by using sulfates. The extractant capacities were in the order: di-(2-ethylhexyl) phosphoric acid > versatic acid 10 > alamine 336. The reaction of yttrium with each extractant involved the formation of complex compounds having concentration ratios of 1:3 and 1:1 for versatic acid 10 and di-(2-ethylhexyl) phosphoric acid, respectively. The extraction mode for yttrium and impurities was optimum at pH levels of 0.95 to 2.25 for di-(2-ethylhexyl) phosphoric acid. Any iron in the original leachant could be entirely removed by acidity control.

The effects of mechanical activation upon the properties of waste trichromatic phosphors have been investigated[150], showing that such treatment has noticeable effects upon the microstructure of the phosphors. Increased solicitation breaks up the crystalline network and improves rare-earth extraction-rates during sulphuric-acid leaching. The recovery rates of yttrium, europium and cerium attained about 96.3, 91.1 and 77.3%, respectively, in the case of waste trichromatic phosphors which had been activated for 1h using a ball-mill rotational speed of 550rpm. Without the mechanical activation, the above recovery-rates were 46.7, 42.3 and 31.2%, respectively.

The phosphor powders which result from the crushing and sieving of fluorescent lamps comprise some 31% of rare earths in the form of $Y_{1.9}Eu_{0.1}O_3$, $Al_{11}Tb_{0.33}Ce_{0.67}MgO_{19}$ and $Al_{10.09}Ba_{0.96}Mg_{0.91}O_{17}:Eu^{2+}$. Direct leaching and mechanically-assisted leaching are unable to recover cerium and terbium values from the $Al_{11}Tb_{0.33}Ce_{0.67}MgO_{19}$ phase. Heat-treatment with NaOH has been found[151] to be successful in dissociating cerium and terbium by replacing rare-earth ions with Na^+ ion so as to form rare-earth oxides and water-soluble $NaAlO_2$. Yttrium, europium, cerium and terbium could be recovered from heat-treated solid by using a 2-step leaching process, followed by recovery from the leachant via oxalic acid precipitation. An extraction rate of better than 95% was attained following treatment at 400C with 150wt% NaOH for 1h. Yttrium- and europium-containing phases did not take part in the heat-treatment, while cerium and terbium phases underwent solid-state reaction with NaOH via a diffusion-limited process with an activation energy of 41.5kJ/mol. Some 15g of mixed oxide with a purity greater than 95% could be recovered from 100 discarded lamps, and this comprised 79% yttrium, 7% europium, 5% cerium and 4% terbium. Microwave-treatment of phosphor and 50wt% NaOH yielded some 42% of yttrium, 100% of europium, 65% of cerium and 70% of terbium recovery within 5min. About 9g of rare-earth oxides and 5g of cerium-enriched leach residue were recovered, using microwave-treatment, within 5min. A pyrolysis system, with heat-ramping ability, made it possible[152] to relate residue data to the temperature ranges required to ensure total mercury desorption. The major disadvantage of such heat-treatments was the amount of mercury that was absorbed from the residue by the glass matrix; ranging from 23.4 to 39.1% of the sample. It was estimated that 70% of the mercury was recovered at 437C. Again considering the waste phosphor phases, $Y_2O_3:Eu^{3+}$, $BaMgAl_{10}O_{17}:Eu^{2+}$, $CeMgAl_{11}O_{19}:Tb^{3+}$ and $LaPO_4:Ce^{3+}Tb^{3+}$, the thermal decomposition activation energies for sulfation reactions were estimated to be 652.9, 375.9 and 409.5kJ/mol at acid dosages of 0.7, 0.85, and 1ml/g, respectively, at 231 to 308C. Microwave-baking (800W) for 3min at an acid ratio of 1ml/g led to 82.5% overall rare-earth dissolution. This included 93.6% terbium, 39.6% lanthanum and about 100% europium and yttrium dissolution. Cerium dissolution was negligible under the above conditions. The dissociation of $LaPO_4:Ce^{3+}Tb^{3+}$ governed overall rare-earth dissolution during baking. The lanthanum and cerium contents interacted with phosphoric acid at the phosphor surface so as to form the polyphosphates, $LaPO_4$ and $CePO_4$; thus impairing the dissolution efficiency. Partially reacted Ce^{III} was oxidized to stable Ce^{IV}, leading to its accumulation in the leach residue. The 184g of phosphors from 100 lamps here yielded 73g of 98% pure Y-Eu-Tb oxides.

A 3-step process for rare-earth recovery involved solid-state chlorination, leaching using a pH of 3 and solvent extraction[153]. The solid-state chlorination was the key step in separating rare earths from residues and involved dry HCl gas which was produced by the thermal decomposition of solid NH_4Cl. The digestion step was optimized by a liquid/solid ratio of 40, leading to a 20% reduction in the use of water. Yttrium and europium were recovered separately from solution by using a 4-stage cross-flow solvent extraction process which combined Cyanex 923 and Cyanex 572 and yielded initial purities of at least 94%. Some 95.7% of the yttrium and 92.2% of the europium were selectively recovered at 295.9C within 67min, using a NH_4Cl/solid ratio of 1.27g/g[154].

Rare earths can be recycled from mercury-containing scrap fluorescent lamps by solid-state chlorination[155] using NH_4Cl. The lamps are typically rich in lanthanum, cerium, terbium, gadolinium … and especially yttrium and europium. Mixing with NH_4Cl, and heating to the latter's decomposition temperature, selectively converts yttrium and europium into their chlorides, with high yields. The degree of selectivity and the yield depend upon the temperature and the NH_4Cl/scrap ratio.

A process for the recovery of rare earths involved the selective separation of 3 phosphor fractions by using methanesulfonic acid, an environmentally friendly acid, as a lixiviant[156]. The halophosphate phosphor was first selectively leached (10l/kg) by using pure methanesulfonic acid at 25C for 2h. The $Y_2O_3:Eu^{3+}$ phosphor was then selectively leached using dilute methanesulfonic acid at 80C for 2h. The remaining phosphor was finally leached using pure methanesulfonic acid at 180C for 6h. The yttrium-rich and lanthanum-rich leachants were purified via solvent extraction using bis(2-ethylhexyl)phosphoric acid, followed by stripping with oxalic acid.

A hydrometallurgical approach was initially used[157] to recycle rare earths from fluorescent-lamp waste. Leaching of metals from the waste was performed by using nitric, hydrochloric, sulfuric and methane sulphonic acid solutions. The separation of rare earths from nitric acid media was achieved by means of solvent extraction, and experiments were carried out on a mixture of trialkylphosphine oxides. The separation of heavier (terbium, europium, gadolinium) rare earths and yttrium from lighter (cerium, lanthanum) rare earths was possible due to their larger separation factors. The selective stripping of rare earths from iron and mercury was easy when using 4M hydrochloric acid. Further recovery of the extracted iron and mercury, using oxalic or nitric acid solutions, permitted re-use of the organic phase. The thermodynamic aspects of yttrium leaching from fluorescent lamps in sulfuric acid, hydrochloric acid and sodium hydroxide have been analyzed[158], and chemical reactions were proposed for the leaching of yttrium oxide and calcium and phosphor compounds in various media. Sulfuric and hydrochloric acids were concluded to be the most suitable for phosphor treatment. A hydrometallurgical method[159] for improving the recovery efficiency of rare earths from $LaPO_4:Ce^{3+}$, Tb^{3+} green phosphors involved using mechanical pre-treatment before leaching. By applying intense friction, the leaching yields of rare earths were increased from 0.9 to 81% at room temperature; due to a change in activation energy. The activation energy was calculated to have been decreased from 68 to 1.4kJ/mol. The difference was attributed to physicochemical changes, including structural decomposition, a specific surface area increase and particle size reduction. Following the sequential removal of the halophosphate phosphor and the red phosphor, 99.0, 87.3 and 86.3% of the lanthanum, cerium and terbium in the $LaPO_4:Ce^{3+},Tb^{3+}$ phosphor could be dissolved. Another recycling process for lamp phosphor waste was based[160] upon the use of the functionalized ionic liquid, betainium bis(trifluoromethylsulfonyl)imide. This permitted selective dissolution of the red phosphor, $Y_2O_3:Eu^{3+}$, without leaching out other constituents. The latter phosphor contains 80wt% of the rare earths sought even though it makes up only 20wt% of the phosphor waste. Rival hydrometallurgical processes leached out the non-valuable halophosphate phosphor, $(Sr,Ca)_{10}(PO_4)_6(Cl,F)_2:Sb^{3+},Mn^{2+}$, while attempting to dissolve the $Y_2O_3:Eu^{3+}$. Europium coordination and Eu^{III}/Eu^{II} electrochemical behavior have been studied[161] as a function of the water content of 1-ethyl-3-methyl imidazolium bis(trifluoromethylsulfonyl) imide. Under anhydrous conditions, Eu^{III} and Eu^{II} were complexed so as to form Eu−O and Eu−(N,O) bonds with the anion sulfoxide function and nitrogen atoms, respectively. This resulted in a greater stability of Eu^{II} and in a quasi-reversible oxidation–reduction, with a potential of 0.18V versus the ferrocenium/ferrocene couple. With increasing water content, increasing incorporation of water into the Eu^{III} coordination sphere occurred, leading to reversible oxidation–reduction reactions and to a decrease in the stability of the +II oxidation state. Because the halophosphate phosphor comprises up to 50wt% of the lamp waste, this can complicate later solvent extraction. The dissolved yttrium and europium can be recovered by using a stoichiometric amount of solid oxalic acid or by bringing the ionic liquid into contact with a hydrochloric acid solution. Both choices regenerate the ionic liquid, but the use of oxalic acid involves no loss of ionic liquid to the water phase and the yttrium/europium oxalate can be calcined so as to re-constitute the red $Y_2O_3:Eu^{3+}$ phosphor at a purity of better than 99.9wt%. An early proposed method[162] for the separation of ultra-fine fluorescent particles (red, green, blue) before recycling involved a two-step process in which each step consisted of a liquid-liquid extraction that was based upon two (non-polar, polar) organic solvents which created two phases, plus a surfactant which controlled the wettability of the powders. In the first step, green powder migrated towards a non-polar phase such as n-heptane and remained at the interface of the two solvents. The other components precipitated in the polar phase. During the second step, blue powder migrated towards a non-polar phase and remained at the interface of two solvents while red powder precipitated in the polar phase. The recovery of each separated powder was greater than 90%. A new cost-effective recycling method has been proposed[163], for the recovery of rare earths from waste-lamp fluorescent powder, which involves the sequential digestion of two phosphor components and the treatment of their leachates separately under given hydrometallurgical conditions. The phosphors were useful sources of yttrium, europium, terbium, lanthanum, cerium and gadolinium. The leaching of fluorescent powder led to a better than 95% recovery of europium, lanthanum, cerium, yttrium and terbium. Recycling to better than 99% purity was possible for yttrium, europium and terbium within 1, 25 and 55 liquid-liquid extraction stages, respectively. A comparative study[164] of the recycling rare earths from waste phosphors concentrated on the leaching rates of traditional and dual dissolution by hydrochloric acid. The red phosphor, $(Y_{0.95}Eu_{0.05})_2O_3$, in the waste was dissolved during the first stage of leaching while the green phosphor, $(Ce_{0.67}Tb_{0.33})MgAl_{11}O_{19}$, and the blue phosphor, $(Ba_{0.9}Eu_{0.1})MgAl_{10}O_{17}$ - mixed with caustic soda - were obtained by sintering and excess caustic soda and $NaAlO_2$ were removed by washing. Insoluble matter was leached out with hydrochloric acid, followed by solvent extraction and precipitation. The total leaching rate of the rare earths using the dual method was 94.6%; much higher than the 42.08% rate possible when using the traditional method. The leaching rates of yttrium, europium, cerium and terbium attained 94.6, 99.05, 71.45 and 76.22%, respectively. Red phosphors, $Y_2O_3:Eu^{3+}$, have been specifically chosen[165] for evaluating the recovery potential of rare earths. The rare earth liquor arising from a soft leaching process has been precipitated by adding oxalic acid, followed by calcining so as to obtain the rare earths in oxide form. Cyanex 572, di-(2-ethylhexyl) phosphoric acid and ionic liquids (Primene 81R,·Cyanex 572L, Primene 81R, di-(2-ethylhexyl) phosphoric acid) were used to investigate the efficiency of rare-earth separation in chloride media. Yttrium, europium and cerium were recovered individually by using a four-stage cross-flow solvent extraction process involving Primene 81R·D2EHPA and Primene 81R·Cyanex 572 as extractants. The above rare earths were recovered in purities of better than 99.9%, while 4mol/l of hydrochloric acid was used to recover yttrium and europium from the organic phases. Cerium and terbium tend to be more difficult to recover due to the spinel structures in which they reside. Mechanical activation has thus been used[166] to pre-treat the scrap phosphors. The rare-earth recovery-rate rapidly increased with increasing milling rotation-rate and activation time. Under optimum conditions, the leaching rates of cerium and terbium attained 85.0 and 89.8%, respectively, while the total rare-earth recovery-rate attained 95.2%. This showed that alkali mechanical activation can destroy spinel structures, leading to easy dissolution of the rare earths in acid solutions and markedly improving the leaching of cerium and terbium.

Fluorescent powders which result from waste lamp treatment can, in some cases be re-used without further purification or separation, to produce $Y_2Al_5O_{12}$ doped with cerium. New phosphors which are obtained in this way have the same crystal structure as that of commercial samples, together with comparable optical properties. As an example[167], the cerium-related emission efficiency had a quantum yield of about 0.75 when excited at 450nm, in good agreement with data on the commercial powders used for white-light light-emitting diodes.

Hard disk drives

The most attractive feature of the drives, with regard to rare-earth recycling, is the magnet content. As well as neodymium, the neodymium–ion–boron permanent magnets which are used in these drives can include praseodymium, dysprosium and terbium. The magnetic material can be liberated from the hard drive by preferential degradation of the brittle magnet material. This process can recover more than 95% of the magnet material[168]. The process also yields stainless steel, aluminium, nickel alloy and carbon steel as by-products. In one technique, the magnets were ground and screened; showing that a $Nd_2Fe_{14}B$ tetragonal phase was the predominant constituent, with the magnet composition comprising 21.5wt% of neodymium and 65.1wt% of iron[169]. This neodymium content was higher than those found in neodymium ores. The Nd-Fe-B permanent magnets collected from hard disc drives consisted of about 28% of rare earths together with 65% of iron in the form of $Nd_2Fe_{14}B$. The magnets were demagnetized and crushed and, within 1.2min of microwave exposure, appreciable oxidation and a temperature of about 600C was established[170]. Metal recovery from the microwaved product was performed by leaching and precipitation. The iron was recovered in the form of metallic iron and iron oxide in the leach residue. The very rapid process led to 56% recovery of neodymium and dysprosium oxides having a purity better than 98%.

A simple approach[171] to the recovery of rare earths from ultrafine Nd-Fe-B scrap is 1-step precipitation in hydrofluoric acid. Rare-earth precipitation is much faster than iron-leaching, thus permitting rare-earth separation following complete iron dissolution.

Rare earths and iron can be recovered[172] from Nd-Fe-B scrap by co-precipitation with urea, yielding nano-sized rare-earth and iron oxide particles and micro-sized Fe_2O_3 powders. The rare-earth recovery first increases and then decreases, while iron recovery always increases, as the pH level is increased from 1 to 6. There were some differences among 4 rare-earth elements with regard to these changes. The recovery yields of 4 rare earths remains above 90% at 60 to 70C, with a continuous temperature-related increase. The optimum precipitation parameters are a pH level of 3 and a temperature of 65C. Recovery yields of up to 94.92 and 100.94% are obtained for iron and rare earths, respectively, with a total recovery yield of up to 98.81% and a purity of up to 98.86%.

When recycling Nd-Fe-B permanent magnet scrap, free carbon and carbides at grain boundaries are harmful and have to be removed[173]. The carbon content has to be lowered to an acceptable level before melting, and decarburization by oxidation in air at above 1273K permits the reduction - down to 300ppm[wt] - not only of free carbon but also of the grain-boundary carbides. The decarburization and de-oxidation of scrap Nd-Fe-B magnets has been investigated[174] because carbon and oxygen impair the magnetic properties. It was found that the carbon content decreased by less than 0.001% upon heating in air. Iron oxides were first reduced by heating in hydrogen and rare-earth oxides were then removed by calcium-reduction and leaching. The heating pattern, during the calcium reduction and leaching of a mixture of calcium compounds and Nd-Fe-B alloy powder, greatly affected the oxygen content of recycled material.

On the other hand, a concomitant increase in the oxygen content is inevitable during the high-temperature oxidation. A further step is therefore to reduce any iron oxide content by heating in a hydrogen atmosphere at 1273K. As a final step, oxygen combined with rare earths is removed by using calcium; with its strong affinity for oxygen. That is, powdered scraps of these magnets are de-oxidized by using calcium vapour and liquid or $CaCl_2$ melts at 1223 to 1273K[175]. Because the scrap reacts with acidic solutions during leaching of the by-product, CaO, methods for the removal of excess calcium and CaO have been studied. When the pH level is maintained at about 8 during several leaching treatments, the dissolution of rare earths can be minimized. Calcium-vapor de-oxidation cannot supply enough calcium through the CaO layer. When a large amount of $CaCl_2$ is added, the dissolution rate of CaO in aqueous solutions increases but some of the rare-earth component is lost. De-oxidation by calcium liquid and 5wt%$CaCl_2$, and subsequent leaching in distilled water with a pH level greater than 8, gives better results such that the oxygen level, even in heavily oxidized scrap, could be decreased to 0.7wt%. An economic decarburization method was considered[176] in which only the carbon sources in scrap neodymium magnets were to be decarburized, without generating iron oxide. Ground magnet sludge could be decarburized to less than 0.03wt%, without generating any iron oxide, by heating at above 1073K under a pressure of less than 5.32×10^{-2}Pa. The amount of oxygen in the decarburized powder was then about 8wt%. Nd-Fe-B permanent magnets have been subjected to oxidation (1000C, 1h) followed by carbothermal reduction (1450C, 1.5h) using carbon, produced from scrap tyres, as a reducing agent[177]. Iron-based metal and rare earth oxide phases were then separated. Rare earths were the main components of the oxide phase, and did not remain in the iron-based metal phase.

Note that magnet recycling need not always involve complete separation of the rare earths. Sintered Nd-Fe-B magnets have been recycled[178] via the hydrogen processing of scrap. Hydrogenated powder was milled and sieved to furnish a range of particle sizes, and the resultant powder was re-sintered at 1060C in vacuum so as to give new magnets. Those which were produced using the smaller particles exhibited an increase in remanence, density and maximum energy product as compared with those made from larger particles. In both cases, the re-sintered magnets exhibited a decrease in density and magnetic performance as compared with those of the starting material. This could be overcome by adding $NdH_{2.7}$ to the recovered hydrogenated Nd-Fe-B powder in order to promote liquid phase sintering. By blending with the hydride, the coercivity of the recycled magnets was recovered and could even exceed that of the starting material. Waste sintered Nd-Fe-B magnets have been recycled[179] by doping with $(Nd_{20}Dy_{80})_{76}Co_{20}Cu_3Fe$ alloy powder. Additions of up to 2.0wt% of the latter had little effect upon the remanence of the recycled magnet but, after adding more than 2.0wt%, the remanence began to decrease. The coercivity of the recycled magnets gradually increased as more additive was used. As compared with the original waste sintered material, samples which were recycled using 2.0wt% of the additive recovered 97.5, 92.4 and 93.1% of the remanence, coercivity and maximum energy product, respectively. The volume fraction of neodymium-rich phase attained a maximum value of 7.4vol% at 2.0wt% of the additive; almost equal to that of the original scrap magnet. The average grain size of the recycled magnets was always greater than that of the original scrap magnet. Scrap Nd-Fe-B magnets have been induction-melted, supplemented with a small percentage of virgin elements and hydrided[180]. The resultant powder was then ball-milled, and isopressed into $Nd_2Fe_{14}B$ sintered magnets. These exhibited a remanence of 11.72kG, an intrinsic coercivity of 14.35kOe and a maximum energy product of 32MGOe. The process could also be applied to scrap magnets of Sm_2Co_{17} type. In this case, the remanence was 10.67kG, the intrinsic coercivity was 11.9kOe and the maximum energy product was 27.1MGOe. The recycled Nd-Fe-B and Sm-Co magnets exhibited properties which were almost equivalent to those of commercial magnets of the same type. An environmentally friendly process has been based[181] upon low-viscosity hydrophobic deep eutectic solvents for the recovery of Sm^{III} from waste SmCo magnets. The solvents, based upon dodecanol and tri-n-octylphosphine oxide were first used to extract Fe^{III} and Sm^{III} from SmCo-magnet leachate. The Cu^{II} and Co^{II} in the raffinate were then separated by using eutectics which were based upon decanoic acid:lauric acid (2:1). Following 2-stage extraction by using eutectics based upon dodecanol:tri-n-octylphosphine-oxide (2:1), more than 99% of the Fe^{III} and Sm^{III} was extracted into the organic phase. These ions could be stripped by using 1.5mol/l $H_2C_2O_4$. The resultant $Fe_2(C_2O_4)_3$ solution and $Sm_2(C_2O_4)_3$ precipitate could be separated by filtration. The final purity of the Sm^{III} was at least 98%, and its recovery rate was 99%.

The recovery of samarium and neodymium from scrap rare earth magnets containing some 30% of samarium or neodymium and 50 to 60% of cobalt or iron has been achieved[182] via the fractional crystallization of their sulfates. When H_2SO_4 was added to a HNO_3 solution containing samarium and cobalt, the solubility of samarium decreased and samarium sulfate hydrate, $Sm_2(SO_4)_3 \bullet 8H_2O$, was precipitated preferentially. Samarium sulfate hydrate of 96.5% purity, with a recovery rate of 87.1%, was obtained from $SmCo_5$ magnet scrap by such fractional crystallization. Good results were also obtained when starting with $Sm_2(Co,Fe,Cu,Zr)_{17}$. The solubility of $Nd_2(SO_4)_3$ in H_2SO_4 solution decreased appreciably upon adding a small amount of ethanol, while the solubilities of $FeSO_4$ and $Fe_2(SO_4)_3$ decreased slightly. The addition of ethanol was effective in the recovery of neodymium, by fractional crystallization, from H_2SO_4-HNO_3 solutions which contained neodymium and iron. Neodymium sulfate hydrate, $Nd_2(SO_4)_3 \bullet 8H_2O$, of 96.8% purity, at a recovery-rate of 97.1%, was obtained from scrap Nd-Fe-B magnets by fractional crystallization upon adding ethanol as well as H_2SO_4.

Scrap nickel-coated Nd-Fe-B sintered magnets have been recycled[183] by melt-spinning. The oxygen content of recycled magnet powder was lower (less than 0.1wt%) than that of the waste sintered magnets. The powder which possessed the best magnetic properties: remanence of 0.78T, intrinsic coercivity of 0.72MA/m and maximum energy product of 87.8kJ/m^3, was obtained when the spinning velocity was 16m/s. The magnetic properties of bonded magnets which were prepared from the above powder were a remanence of 0.69T, a coercivity of 0.70MA/m and a maximum energy product of 71.0kJ/m^3. These were essentially the same as those which were obtained by using commercial powder. A remanence of 0.74T, an intrinsic coercivity of 0.94MA/m and a maximum energy product of 86.5kJ/m^3 were obtained[184] when amorphous-like material, prepared using a spinning velocity of 25m/s, was annealed for 180s at 973K in argon. These properties were further improved by adding iron to the original scrap, as this increased the intrinsic magnetization. The properties of bonded magnets which were prepared from the above powders at 1GPa were a remanence of 0.69T, an intrinsic coercivity of 0.81MA/m and a maximum energy product of 73.8kJ/m^3; again very similar to those which resulted from using commercially available powder. Polymer-bonded magnets have been rapidly pulverized[185] into composite powder which contained Nd-Fe-B particles and polymer binder by milling at cryogenic temperatures. Recycled bonded magnets which were made by warm compaction of the coarse ground cryomilled composite powders and nylon particles exhibited improved magnetic properties and densities. The remanence and saturation magnetization were increased by 4 and 6.5%, respectively, due to the increased density. The coercivity and energy product were preserved from the original material. Starting with neodymium, praseodymium and dysprosium which had been recovered[186,187], from scrap Nd-Fe-B magnets, urea-based homogeneous precipitation at low temperatures was used to produce oxalates with diameters of 40 to 50nm and specific surface areas of 60m^2/g which were then used to synthesize rare-earth oxide nano-sheets having dimensions of 1μm x 0.5μm x 17nm by thermal degradation at 700C. Strong X-ray diffraction peaks were attributed to the cubic phase of Nd_2O_3 while small peaks corresponded to the cubic phases of $NdPrO_3$ and Pr_2O_3. This indicated that Nd^{3+} in the Nd_2O_3 host lattice was replaced by Pr^{3+} and Dy^{3+}.

A 3-step process was developed[188] for recycling the neodymium from Nd-Fe-B magnet particles in the ferrous fraction of shredded waste electrical and electronic equipment. Upgraded ferrous waste was first oxidized by means of water corrosion, and then leached with dilute H_2SO_4 so as to extract neodymium and other non-ferrous elements selectively. The leachant was finally treated with Na_2SO_4 so as to precipitate neodymium as the double sulphate, $(Nd,Na)(SO_4)_2$. The oxidation process oxidized 93% of the metallic iron to $Fe(OH)_3$, and leaching dissolved between 70 and 99% of the neodymium; depending upon the temperature and the liquid/solid ratio. The precipitation step recovered 92% of the leached neodymium, and the purity of the precipitates depended upon the pH at which the precipitation took place: a pH level of less than 0.5 was required in order to prevent iron contamination and a negative pH level reduce calcium contamination to below 1wt%.

An environmentally friendly process for the acid-free leaching of rare earths and cobalt has been developed[189] for the treatment of magnet-containing electronic wastes. The use of copper salts for oxidative dissolution permits the selective dissolution of the relevant metals. It also minimizes the use of strong acids. The process additionally allows the copper content of the salts to be re-used. In the case of Nd-Fe-B magnets, the basic room-temperature reaction is,

$$2\mathfrak{R}_2Fe_{14}B + 34Cu^{2+} + 10.5O_2 \rightarrow 4\mathfrak{R}^{3+} + 28Fe^{2+} + Cu_3(BO_3)_2 + 15Cu_2O + Cu^0$$

In the case of Sm-Co magnets, the basic room-temperature reaction is,

$$2SmCo_5 + 13Cu^{2+} + O_2 \rightarrow 2Sm^{3+} 10Co^{2+} + 2Cu_2O + 9Cu^0$$

The closed-loop recycling of Sm-Co waste via acid-free leaching was efficient with regard to both samarium and cobalt recovery.

Another twist is firstly to embrittle magnetic material by using liquid nitrogen, before dissolution in an acidic solution and the addition of oxalic acid solution[190]. Even underwater explosion has been used[191] for the liberation of neodymium magnets from motors. Dissolution leads to the formation of the hydrated oxalate, $Nd_2(C_2O_4)_3 \bullet 10H_2O$, having a purity of better than 99%. Thermal decomposition then transforms this powder into neodymium oxide.

Clays in the montmorillonite group have been used[192] as sorbents for the removal of rare earths from the leachant of scrap Nd-Fe-B magnets. The clays were able to capture and release lanthanum and neodymium ions via an ion exchange mechanism. The best total efficiencies, of 50% for capture and 70% for release, were obtained when uptake and release were performed at pH levels of 5 and 1, respectively. For typical scrap leachant the uptake was about 40% but the release efficiency decreased from 80 to 5% in going from a single-ion system to a real system. Two solid matrices were later considered: a pristine montmorillonite clay and montmorillonite clay intercalated with pentaethylenhexamine. The capture ability was tested with respect to single-ion lanthanum, neodymium and yttrium solutions or a multi-element solution which contained all 3 ions[193]. In both cases, at lower initial concentrations, the ions were captured to a similar degree. At higher concentrations, the pure clay had a high total uptake of lanthanum ions and this was attributed due to surface interactions. The modified clay interacted preferentially with neodymium and yttrium, and this was attributed to ion coordination with amino groups. The overall capture behaviour was related to the physicochemical properties of the ions as well as the ionic radius.

Hydrometallurgical recovery of neodymium typically consists of pre-treatment, chemical leaching and metal precipitation. A life-cycle assessment of Nd-Fe-B magnet production was compared[194] with one for Nd-Fe-B magnet production from bastnäsite/monazite ores using traditional sintering methods. This revealed that, from both the economical and

Materials Research Forum LLC
https://doi.org/10.21741/9781644901793

environment perspectives, magnet production from recycled neodymium was superior to virgin magnet manufacture. Scaling-up of neodymium recovery reduced the environmental impact of Nd-Fe-B magnet production by up to 65%, and reduced the production costs from 8.55 to 3.98$/kg. Recycled magnets compete well with those made from primary materials, in terms of magnetic performance and production costs, and are clearly superior in terms of environmental protection[195]. Physical and hydrometallurgical methods such as de-magnetization, grinding, screening, leaching and precipitation have been used[196] to recover neodymium and dysprosium from scrap Nd-Fe-B magnets. Inorganic acids (HCl, HNO_3, H_2SO_4, aqua regia) and organic acids (acetic, oxalic) were used to leach the magnets. Iron was eliminated by precipitation as hydroxide. The extraction efficiency of di-2-ethylhexyl phosphoric acid was found to be higher than that of trihexyltetradecylphosphonium chloride, following the leaching stage. The ionic liquid, trihexyltetradecylphosphonium trichloride, [P666,14][Cl_3], can safely store chlorine gas in the form of the trichloride anion. It can be used[197] as an oxidizing solvent for the recovery of metals from scrap SmCo magnets. The maximum capacity of [P666,14][Cl_3] for the magnets is 71mg/g in the presence of an extra source of chloride ions, and the maximum loading can be reached within 3h at 50C. Four stripping steps can remove all metal from the loaded ionic liquid when NaCl solution (3mol/l), water and ammonia solution (3mol/l) are used consecutively as stripping solvents. The regenerated ionic liquid retains much of its dissolution ability. Pure [P666,14][Cl_3], or a mixture with [P666,14]Cl, could dissolve[198] Nd-Fe-B magnets when the solid/liquid ratio was less than a certain threshold value that depended upon the volume percentage of [P666,14][Cl_3]. Increasing the temperature from 25 to 50C markedly increased the dissolution rate, but the dissolution efficiency was only slightly increased. The volume percentage of [P666,14][Cl_3] in [P666,14]Cl had a positive effect upon the dissolution efficiency. Rare-earth and transition metals could again be selectively removed in 2 sequential stripping steps, using 3mol/l NaCl aqueous solution followed by more than 2mol/l of aqueous ammonia solution. Even regenerated [P666,14][Cl_3] exhibited a similar dissolution efficiency to that of fresh liquid.

A laboratory-scale process has been described[199] for the recycling of rare earths from ℜ-iron-boron permanent magnets by using chlorine gas at 673K for 2h. This treatment could be applied without prior demagnetization, crushing or milling. Following treatment at 673K, a clinker powder was found which consisted of rare-earth chlorides plus small amounts of other metals. Any haematite or iron oxychloride had sublimated. It was assumed that the rare-earth chlorides in the clinker could be easily reduced to metal by electrolysis in relatively low-temperature eutectic melts, or by reduction in mixtures with alkali or alkaline earth metals.

A novel environmentally friendly hydrometallurgical route for the efficient recovery of rare earths during electrochemical leaching using sulfuric and oxalic acids permits, with suitable adjustment of the electrolyte and operating conditions, the effective separation of the elements[200]. A compact layer of rare earth oxalates, having a purity of up to 93%, appears on the cathode while iron remains in solution or as a solid residue for further processing. The cathodic deposition of the rare earths can be attributed to electrostatic attraction of their oxalate particles to the cathode. This effect could be exploited for the selective recovery of individual rare earths from scrap magnets. Using other techniques[201], organophosphoric acid ligands (bis-2-ethylhexyl phosphoric acid, 2-ethylhexylphosphonic acid mono-2-ethylhexyl ester) exhibit a high selectivity for heavier rare earths. Molecular dynamics simulations have been used to clarify the structure of neodymium- and dysprosium-bis-2-ethylhexyl phosphoric acid complexes in vacuum, aqueous and organic phases. This showed that the selectivity of bis-2-ethylhexyl phosphoric acid for dysprosium arises from a favorable differential stabilization of the complex in solvents phases which is caused by structural features of the dysprosium-bis-2-ethylhexyl phosphoric acid complex. Solvent extraction experiments, performed using a 4:1 mixture of neodymium and dysprosium ions in chloride media with n-heptane diluents, showed that - although the dysprosium concentration was 4 times smaller - the selectivity of bis-2-ethylhexyl phosphoric acid toward dysprosium could be exploited so as to obtain improved separation in a 2-step process by first extracting the dysprosium at a low pH level and minimal contents of bis-2-ethylhexyl phosphoric acid and then extracting the neodymium at higher pH levels. A so-called aeriometallurgical process has been proposed[202] for the recycling of neodymium, praseodymium and dysprosium from scrap Nd-Fe-B magnets. This process uses super-critical CO_2, a non-toxic inert and abundant solvent, together with a tributyl-phosphate-nitric acid chelating agent and 2wt% of methanol as a co-solvent. Some 94% extraction of neodymium, 91% of praseodymium and 98% of dysprosium is possible, with only 62% extraction of iron and minimal waste. The metal-ion charge has an important effect upon the extraction efficiency. The extraction proceeds via corrosion of the surface layers of the magnet particles. An earlier recycling process for microwave-roasted Nd-Fe-B magnets was based[203] upon the carboxyl-functionalized ionic liquid, betainium bis(trifluoromethylsulfonyl)imide. By exploiting the thermomorphic properties of an aqueous mixture of the latter, a combined leaching and extraction combination could be used. A change from an homogeneous system during leaching at 80C, to a two-phase system at room temperature, causes dissolved metal ions to redistribute themselves between the two phases. Neodymium, dysprosium and cobalt can thereby be efficiently separated from iron. Further processing used oxalic acid to precipitate rare earth and cobalt ions, while transferring iron ions from the ionic liquid to the water phase as a soluble oxalate complex. Any cobalt, when present, was removed by treating the mixed oxalate precipitate with aqueous ammonia. The remaining rare-earth oxalate was calcined so as to produce the corresponding rare-earth oxide with 99.9% purity. The ionic liquid was then regenerated, and contamination of the water phase was avoided by salting-out the ionic liquid using Na_2SO_4. In a similar process[204], rare earths were selectively dissolved from crushed and roasted Nd-Fe-B magnet waste by using a minimal amount of acid (HCl, HNO_3), purified by solvent extraction and again precipitated as pure oxalates; leaving iron behind. Any remaining cobalt, copper and manganese was dissolved in the ionic liquid, trihexyl(tetradecyl)phosphonium chloride. Rare earths were precipitated by adding oxalic acid and the precipitate was again calcined to give oxides. The dissolution of oxides used 4 molar equivalents less acid to dissolve all of the rare earths than did the dissolution of non-roasted magnet scrap, with the added advantage that iron was already removed. The hydrochloric acid which was a by-product of oxalate precipitation could be re-used for further selective leaching. The overall recycling process thus consumed only air, water, oxalic acid and electricity. In a further use[205] of trihexyl(tetradecyl)phosphonium chloride, transition metals were removed from neodymium-iron-boron or samarium-cobalt permanent magnets. The highest distribution ratios for cobalt and iron were found for 8.5 and 9M HCl, and the concentrations of neodymium and samarium in the ionic liquid were below 0.5mg/l; even when the initial concentrations was 45g/l. The separation factors of neodymium/iron and samarium/cobalt were 5.0 x 10^6 and 8.0 x 10^5, respectively. The percentage extraction of iron was greater than 99.98% when the ionic liquids contained 70g/l of iron. The viscosity of ionic liquid which contained the tetrachloroferrate complex, $[FeCl_4]^-$, was lower and less dependent upon the initial concentration, than it was in the case of the tetrachlorocobaltate anion, $[CoCl_4]^{2-}$. Following extraction, cobalt could be very easily removed from the ionic liquid by water. Due to the very high distribution ratio, iron could be stripped only by forming a water-soluble iron complex with ethylenediaminetetraacetic acid. Another solvent extraction process[206] involved the ionic liquid, trihexyl(tetradecyl) phosphonium nitrate, which separated rare earths from nickel or cobalt. The ionic liquid could be prepared by a simple metathesis from trihexyl(tetradecyl)phosphonium chloride. Extraction was aided by the salting-out effect of a highly concentrated metal-nitrate aqueous phase. When starting solutions containing 164g/l of cobalt and 84g/l of samarium, or 251g/l of nickel and 61g/l of lanthanum were tested, percentage extractions of better than 99% were possible for the rare earths. Following further scrubbing, the purity of the rare earth in the ionic liquid was 99.9%. Complete regeneration of the ionic liquid was possible by using only pure water. A high viscosity and sluggish mass-transfer explained why non-fluorinated ionic liquids generally had to be diluted with conventional hydrophobic solvents such as kerosene, toluene or chloroform. During the extraction of samarium and lanthanum, differing anionic complexes were formed: lanthanum was extracted at maximum loading via the hexakis anionic complex, $[La(NO_3)_6]^{3-}$, while samarium was extracted at maximum loading via the pentakis anionic complex, $[Sm(NO_3)_5]^{2-}$. The differing electrical charges on the anions had a marked effect upon the viscosity of the ionic liquid phase. An aqueous two-phase system which consisted of an ionic liquid, tributylmethylammonium nitrate, and $NaNO_3$ solution was designed[207] for the separation of rare earth ions, such as Nd^{III}, from transition-metal ions. Efficient separation of Nd^{III} from Fe^{III}, Ni^{II} and Co^{II} was found to be possible. A similarity of the environment across the liquid/liquid interface was indicated by an ultra-low viscosity and interfacial tension of the ionic-liquid-rich phase, and this was held to be responsible for the good Nd^{III} extraction kinetics. Some 2mol/l of HNO_3 aqueous solution could effectively strip Nd^{III} from the loaded ionic-liquid-rich phase. The extraction performance of the ionic liquid remained essentially unchanged after 5 cycles. So-called deep-eutectic solvents, such as one based upon choline chloride and lactic acid at a molar ratio of 1:2, have been used[208] as an alternative to aqueous solutions for the recovery of metals from Nd-Fe-B magnets. The separation of iron, boron and cobalt from neodymium and dysprosium in the deep-eutectic solvent was achieved by using the ionic liquid, tricaprylmethylammonium thiocyanate (Aliquat 336 SCN, [A336][SCN]), diluted in toluene. The stripping of boron was carried out by using HCl, while ethylenediaminetetra-acetic acid was used for the recovery of iron and cobalt. The separation of neodymium and dysprosium was carried out by using a mixture of trialkylphosphine oxides (Cyanex 923) or bis(2-ethylhexyl)phosphoric acid. On the basis of the distribution ratios and separation factors, Cyanex 923 was judged to be the more effective extractant. Purified dysprosium in the less polar phase was easily recovered by stripping with water. Neodymium in the deep-eutectic solvent was recovered by precipitation stripping with a stoichiometric amount of oxalic acid, with Nd_2O_3 and Dy_2O_3 being recovered in purities of 99.87 and 99.94%, respectively.

A liquid metal extraction process, using molten magnesium at a magnet/magnesium mass ratio of 1:10, has been used[209] to produce neodymium-magnesium alloys from neodymium-based permanent magnets at 900C after 24h. The neodymium content of the alloy was about 4wt%, and this was recovered from the alloy by exploiting the difference in their vapor pressures using vacuum distillation at 800C under 2.67Pa. Neodymium with a purity of more than 99% was recovered after distilling for more than 2h. The liquid metal extraction process, again using liquid magnesium, was applied[210] to the recovery of dysprosium from a rapidly solidified Dy–Fe–B alloy system consisting of $Dy_2Fe_{14}B$ and Dy_6Fe_{23}. Magnesium was chosen because it forms intermetallic compounds with dysprosium but not with iron or boron. The extraction rate increased with increasing temperature, and the maximum dysprosium extraction efficiency was about 74% after 1h at 1000C. As the reaction-time was increased, a maximum extraction efficiency of 95% dysprosium was found after 24h at 900C. The factor which governed the dysprosium extraction ratio at up to 6h was the Dy_6Fe_{23} phase, after which it was mainly the $Dy_2Fe_{14}B$ phase.

Binary chloride mixtures of rare earths have been separated[211] by using a reduction vacuum distillation process. The apparent separation factors were 8.1 for a praseodymium-neodymium chloride mixture and 570 for a neodymium chloride mixture; values which were much higher than those for other solvent-extraction methods. Rare earths in neodymium magnet sludge were extracted by chlorination with $FeCl_2$, and activated carbon was used for de-oxidation. Metallic iron in the sludge could not be chlorinated, because of the instability of iron monochloride. Extracted rare-earth chlorides could be easily separated from iron alloy and excess $FeCl_2$ by vacuum distillation, with 96% of neodymium and 94% of dysprosium in the sludge being extracted as a chloride phase. Vacuum distillation yielded a mixture of neodymium and dysprosium trichlorides of 99.2% purity. The rare-earth chlorides could be converted into their corresponding oxides via pyrohydrolysis and the formation of HCl gas, with the latter being able to chlorinate metallic iron to $FeCl_2$. This process consumed only carbon and water and generated no toxic pollutants.

When recycling scrap Fe-Nd-B magnets using solid-state chlorination[212,213] the latter takes place via the partial decomposition of NH_4Cl at between 225 and 325C. Dry HCl gas reacts with the magnet material so as to form water-soluble metal chlorides which are then dissolved in an acetic acid buffer medium. The maximum rare-earth yield can attain 84.1%, and the process is reduces chemical consumption by 45%. When Sm-Co alloys are chlorinated[214] under the same conditions as Fe-Nd-B magnets, the particles which arise from $SmCo_5$ scrap disintegrate more rapidly than expected during chlorination. This aids the selective chlorination of samarium, and yields can attain 99.7%.

The high-temperature recycling of neodymium magnets has been essayed by using B_2O_3 flux, but the latter leads to a poor-fluidity slag. The B_2O_3 was therefore replaced[215] by $Na_2B_4O_7$ because of its lower viscosity. The Nd_2O_3–$Na_2B_4O_7$ phase diagram at 1460 to 1780K showed that the pseudo–binary system formed an homogeneous melt at between 16 and 20mass%$Na_2B_4O_7$ and at over 55mass%$Na_2B_4O_7$ at 1673K. When neodymium magnets and other scrap was melted together with $Na_2B_4O_7$ in a carbon crucible, neodymium, praseodymium, dysprosium and terbium enriched the slag while iron contributed to the Fe–C alloy phase.

Rare-earth oxide can be removed from scrap magnet alloy by re-melting the scrap, together with a flux comprising LiF-50mol%NdF$_3$ and LiF-25NdF$_3$-25mol%DyF$_3$, at 1503K[216]. Separation of the magnet alloy from the fluoride flux following re-melting is complete, with no alloy left in the flux and no flux in the alloy.

A considerable problem in the recovery of Nd-Fe-B bonded magnet waste is how to remove the epoxy resins completely. Chemical reaction and dissolution have been combined[217] so as to remove the resins by adding aqueous ammonia solution and a mixture of organic solvents. The ammonia could react with the epoxy functional group of the resin so as to generate polyols. Mixtures of alcohol, dimethyl formamide and tetrahydrofuran can dissolve these polyols and residual resin. The resins are largely removed under optimum conditions.

Diffusion dialysis has been suggested[218] for the recovery of neodymium and praseodymium from Nd-Fe-B magnets. Four types of polymer membrane were prepared by blending cellulose tri-acetate and polyethylenimine, with the addition of di-(2-ethylhexyl) phosphoric acid, tridodecylamine, tri-octylamine or tri-octylphosphine oxide. Dilute magnet leachate was used as the feed solution. It was possible to extract up to 15% of the boron within 6h by spontaneous diffusion through a cellulose tri-acetate/polyethylenimine/tri-octylamine membrane. When the membranes were positively charged during operation, rare earths which were present mainly in the form of trivalent cations such as Nd^{3+} and Pr^{3+}, were strongly rejected. A reasonable correlation was found between the membrane's water-uptake and the boron transfer; with boric acid molecules passing more readily through more hydrated membranes. The selectivity between boron and rare earths resulted from an interaction between membrane structure, water-uptake capability and surface charge. It also depended upon the leachate composition. A cellulose tri-acetate/polyethylenimine/tridodecylamine membrane offered the greatest boron/rare-earth selectivity, with factors ranging up to 3706 for neodymium and 140 for praseodymium. The lower selectivity with respect to praseodymium was attributed to the lower Gibbs energy-of-hydration of Pr^{3+} as compared with that of Nd^{3+}. Cellulose nanocrystals, used as a liquid crystal template, and tetra-ethyl orthosilicate, used as a silicon-source precursor drive, were self-assembled into thin-film materials by using a sol-gel method and then calcined to yield carbon-based materials at high temperatures in a nitrogen atmosphere so as to improve the mechanical properties of the films and increase the surface area[219]. Then, 3-aminopropyl tri-ethoxysilane and 1-(2-pyridylazo)-2-naphthol were introduced into the surface of the carbon-based thin-film materials, via chemical grafting, so as to improve the adsorption of rare earth. The adsorption performance of heavy rare-earth ions on carbon-based silicon membranes was better that that of light rare-earth ions. At a pH level of 7, the adsorption efficiency could attain more than 90%. The same membranes could also be used many times. Cellulose has recently been nano-engineered[220] so as to develop a bio-based technology, so-called anionic hairy nanocellulose, for the high-capacity selective removal of neodymium ions from aqueous media. The material consisted of fully-solubilized dicarboxylated cellulose chains and cellulose nanocrystals decorated with dicarboxylated cellulose hairs having a charge density about one order of magnitude higher than that of conventional cellulose nanocrystals. The unique colloidal properties, and especially the polyanionic hairs, permit the removal of about 264mg/g of Nd^{3+} in the nano-adsorbent within seconds. As well as allowing Nd^{3+} removal at initial concentrations greater than 150ppm, the material can be fully neutralized and precipitated at concentrations below 100ppm while retaining its partial colloidal stability. The Nd^{3+} removal can be improved by complementary calcium ion-mediated colloidal bridging.

Laptops

A study was made[221] of the physical and chemical characteristics of the various components of laptops in order to determine the locations of metallic items available for recycling. The recycling possibilities were evaluated in terms of the sequential disassembly, separation sorting of components such as the body (49.8wt%), printed circuit board (9.7wt%), hard disk drive (4.9wt%) and battery (12.4wt%). The printed circuit boards harbored copper (25wt%), tin (5.8wt%), and lead (3.1wt%). Precious metals such as gold and silver could make the recycling economical and were abundant in the integrated circuits, capacitors, resistors and processors. Critical elements such as lithium, cobalt and the target rare earths were found in the batteries and hard disk drives, as described at length above. The amount of material that could be recycled was estimated to range from 36 to 100%. The average laptop was anticipated to contain some 386g of copper (14.45wt%), 49.73g of cobalt (1.86wt%), 346mg of silver, 141.2mg of gold and 650mg of rare earths; especially neodymium and dysprosium.

Mobile phones

The metal fraction associated with the printed circuit board and camera parts can be separated, and pulverized into particles with a size of less than some 2mm[222]. The metal is then dissolved in *aqua regia*, and the pH of the solution is increased to 10.5 by adding NH_4OH. The first precipitate is iron oxide; produced by raising the pH to between 3.1 and 4.2. Copper chloride and rare-earth complexes then appear at a pH of between 5.7 and 7.7 and between 8.3 and 10.5, respectively. An alternative biological method is to add filtrate, with a pH of 7.7, to a metal-reducing bacteria growth medium. After two weeks, rhodochrosite and calcite are precipitated as nano-sized minerals. Tantalum capacitors can be visually distinguished from other components, and neodymium can be

Materials Research Forum LLC
https://doi.org/10.21741/9781644901793

detected in easily separable non-magnetic components with sizes of between 0.5 and 1.5mm. Tantalum-rich powder with a grade of 50% was obtained from the capacitors by leaching followed by an oxidizing heat treatment. Neodymium-rich fractions of between 4 and 14% were obtained[223]. It has been pointed out[224] that phones are subject to the phenomenon of so-called hibernation. This is a reluctance to re-cycle. Most mobile phones are replaced within 3 years, but use of the previous phone as a spare is the main cause of hibernation. The willingness to recycle appears to be inversely proportional to the value of the old phone and proportional to the age of the owner.

Superconductors

So-called high-temperature superconductors can hardly be termed consumer goods but their preparation process may fail or they may still need to be eventually recycled at some point. On the other hand, their essential purity makes them more suitable for re-use than for rare-earth extraction. A study[225] has thus been made of the microstructural and mechanical properties of recycled YBCO-type superconductors. It is found that recycled samples exhibit trapped magnetic fields which are equal to some 70 to 80% of those of as-manufactured material. There is a marked reduction in the porosity, and a simultaneous improvement in the distribution of Y_2BaCuO_5 second-phase inclusions within the microstructure of recycled single-grain YBCO samples. The superconducting properties of recycled samples were generally inferior to those of as-manufactured material, but the flexural strength, hardness and tensile strength of the recycled samples could show improvement: the recycled YBCO had an average flexural strength of 75MPa, and this was almost 50% higher than that of as-manufactured material.

Other electrical and electronic scrap

Multi-step leaching[226] can extract rare earths from the dust which is produced during the industrial shredding of waste electronic and electrical devices[227,228,229]. Double-oxidizing with sulfuric acid first dissolves high percentages of most of the rare earths in the dust. Some 50% of any gold present is then extracted by a second leaching step using 0.25M thiourea in a solid/liquid ratio of 0.2g/70ml and 600rpm. Another study[230] investigated the recovery of rare earths from leachants by using Versatic 10 as the carrier in an organic phase and oxalic acid as the extraction agent. Cerium, lanthanum and yttrium could be recovered in high percentages by using 200mM of Versatic 10, loaded with 100mM of tributyl-phosphate in kerosene at a neutral pH. The use of 750mM of oxalic acid led to the recovery of 7.63 and 13.82mg/kg of lanthanum and yttrium, respectively. The use of Cyanex 572 also permits[231] the efficient recycling of rare earths from scrap while using less acid and alkali than other hydrometallurgy processes. Alkali saponification, acid stripping and oxalic-acid precipitation can also be replaced by a single step involving precipitation-stripping-saponification with ammonium fluoride which contains abundant F^- ions in solutions of weak acidity. The F^- was readily complexed with rare earths on the extractant as a precipitate while NH_4^+ ions bonded with the anions of the extractant. The total chemical consumption could thereby be reduced by over 80%, while waste-water was decreased by more than 90%. The rare-earth recovery rate was also greatly increased. The ammonium fluoride could moreover be re-used, with no impact on the environment.

The interaction between self-supported flower-like nano-$Mg(OH)_2$ and low concentrations of rare earths in waste-water has been investigated[232], showing that more than 99% of the rare earths were successfully taken up by the nano-$Mg(OH)_2$. The rare earths could be collected on the surface of the $Mg(OH)_2$ as metal hydroxide nanoparticles having size of less than 5nm. A method was developed, for further separation of the rare earths and residual hydroxide, which involved varying the pH level of the solution.

A selective recycling system for metal ions has been developed[233] which is based upon homogeneous liquid-liquid extraction using a fluorosurfactant. Sixty-two ions (aluminium, barium, beryllium, calcium, cadmium, chromium, cobalt, copper, gallium, gold, hafnium, indium, iridium, iron, lead, magnesium, manganese, mercury, molybdenum, nickel, niobium, osmium, palladium, platinum, rhenium, rhodium, ruthenium, silver, strontium, tantalum, tin, titanium, tungsten, vanadium, yttrium, zinc, zirconium, … antimony, arsenic, bismuth, boron, germanium, phosphorus, selenium, silicon, tellurium, thallium, … cerium, dysprosium, erbium, europium, gadolinium, holmium, lanthanum, lutetium, neodymium, praseodymium, samarium, scandium, terbium, thulium, ytterbium) were considered. By changing the pH from neutral or alkaline (≥ 6.5) to acidic (<4.0), gallium, zirconium, palladium, silver, platinum and rare earths could be extracted, with greater than 90% efficiency, into a sedimented Zonyl FSA, $CF_3(CF_2)_n(CH_2)_2S(CH_2)_2COOH$ liquid phase (n = 6 to 8). All of the rare earths were extracted in higher percentages; the sedimented phase was maintained by using a filter together with a mixed solution of THF and 1M sodium hydroxide aqueous solution. The Zonyl FSA was filtrated, and rare earths were retained on the filter as hydroxides. The filtrated Zonyl FSA was also re-usable. Three different processes have been proposed[234] for the recycling of rare earths from mine-tailings and shredded electrical and electronic equipment. One process extracted rare earths and phosphorus from apatite in mine-tailings by acid leaching followed by cryogenic crystallization and solvent extraction. This purified both the rare earths and the phosphorus, and recovered 70 to 100% of the rare earths from the apatite and over 99% of the phosphorus. Another low-cost and efficient process recovered neodymium from the ferrous portion of scrap electronic equipment by means of water corrosion, followed by acid leaching and precipitation, leading to an overall neodymium recovery of over 90%. A third process recovered both neodymium and iron from shredded electronic scrap by smelting the latter before leaching to produce metallic iron and neodymium-rich slag. The recovery rates of neodymium and iron were better than 90%; with minimal waste but high energy consumption. Given the prevalence of oxalic use, the recycling of rare-earth waste-water from deposited oxalic acid was investigated[235], showing that the metal-ion content was relatively low while the total organic carbon content was 4661mg/l. Following distillation, the recovery rate of 5mol/l HCl was almost 330ml/l and the HCl of 1mol/l could attain 600ml/l. The oxalate-crystal production rate was 16g/l of waste-water and the purity was greater than 99.5%. A clean and efficient process for the recovery of rare earths from scrap cathode-ray tube phosphors is to use[236] a mixture of sulfuric acid and hydrogen peroxide for the oxidative leaching of the rare earths. The leaching efficiencies of yttrium and europium attain 99% under the optimum leaching conditions (3M H_2SO_4, 4vol% H_2O_2, 55C, 1h). The method also avoids creating any sulfur pollution. The ionic liquid, [OMIm][PF6], and the extractant, Cyanex272, have been used for the separation of the rare earths, the optimum parameters being: 0.2mol/l H_2SO_4, 0.4vol% Cyanex272 in the organic phase, 1200s, 25C. Under these conditions, the extraction efficiencies of yttrium, europium, zinc and aluminium were 99, 87, 8 and 0%, respectively. The separation factor of rare earths with respect to zinc reached 593. The extraction system could be recycled and re-used. The leaching process was controlled by diffusion via the product layer, and the apparent activation energies for yttrium and europium were 75.86 and 77.06kJ/mol, respectively. A cation exchange reaction was suggested to occur between the rare earths and the Cyanex272.

Printed circuit boards are a major part of any electrical or electronic device and comprise a non-conductive substrate, overlaid with copper. They also contain appreciable amounts of nickel, tin, aluminium, gold, silver and rare earths; with one tonne of such boards containing up to 1.5kg of gold and up to 210kg of copper. High-temperature pyrolysis of scrap printed-circuit boards has been carried out[237] (850C, 0.25h) in horizontal resistance and thermal plasma furnaces which imposed various degrees of turbulence. Most of the rare earths were found to be concentrated in a carbonaceous residue, with negligible amounts being recovered in the metallic fraction. Most of the recovered rare earths exhibited a high affinity for refractory oxides, silica and alumina, but little affinity for copper, lead or tin. The rare-earth yield was much higher for plasma-furnace treatment, and revealed that turbulence plays an important role in the dissociation and diffusion of rare earths during pyrolysis. Lanthanum, praseodymium, samarium and yttrium required turbulence for their recovery while neodymium, gadolinium, cerium and dysprosium were relatively easy to dissociate and extract.

Materials Research Forum LLC
https://doi.org/10.21741/9781644901793

Following acid digestion, inductively coupled plasma-mass spectrometry has revealed[238] the presence of rare earths in new and old consumer plastics with various polymeric compositions. X-ray fluorescence spectrometry identified bromine and antimony as markers for the presence of brominated flame retardants and Sb_2O_3. At least one rare earth was detected in 24 samples, with 4 samples indicating total concentrations of up to 8mg/kg. Rare earths were most often detected in samples which contained bromine and antimony at levels which did not support a connection with flame-retarding. Various rare earths were also present in plastics which contained no detectable bromine or antimony. These observations suggested a general tendency towards the rare-earth contamination of plastics. Whether the rare-earth content can here be regarded as a source, or merely a pollutant, is currently undecided.

The use of fluxes to process magnesium alloys leads[239] to the generation of so-called black dross. The latter contains metallic and non-metallic phases, including an appreciable fraction of rare earths. It can be separated, by crushing and screening, into metallic and non-metallic fractions with the non-metallic fraction being treated by water and acid leaching. In the case of water-leaching, NaCl, KCl and $CaCl_2$ are separated out for crystallisation. In the case of acid-leaching, the residue is treated with hydrochloric acid so as to dissolve rare earths such as cerium, lanthanum, neodymium and praseodymium. Selective precipitation using oxalic acid, and solvent extraction using di-(2-ethylhexyl) phosphoric acid, permits up to 92.6% of the rare earths to be recovered from the oxide-salt fraction.

Non-consumer waste

Ash

An analysis has been made[240] of the binding characteristics of rare earths in coal ash. The main components of the coal ash were oxides that were composed of silicon, iron, aluminium and calcium … plus residual carbon. Bottom ash and fly ash contained 185.8 and 179.2mg/kg of rare earths, respectively. Some 85% of the rare earths were present in the residual fraction of both bottom ash and fly ash. The results indicated that the rare earths were strongly bound in both bottom and fly ash, and that very strong acids would be required for their thorough extraction. It was noted that 46.3% of rare earths could be recovered from the waste water that was produced during the processing of coal ash-derived zeolites. A process based upon the ionic liquid, betainium bis(trifluoromethylsulfonyl)imide, has been developed[241] for the preferential extraction of rare earths from coal fly ash. Efficient extraction depended upon the liquid's thermomorphic behavior with respect to water. Upon heating, water and the ionic liquid form a single liquid phase and rare earths are leached from the ash via a proton-exchange mechanism. Upon cooling, the water and liquid separate and the leached elements are partitioned between the two phases. Alkaline pre-treatment greatly improved the rare-earth leaching efficiency in the case of the more recalcitrant ashes. Weathered ash permitted a slightly higher rare-earth leaching efficiency than did non-weathered ash. Regardless of the type of ash, the extraction of scandium was especially efficient. The enrichment of rare earths by concentrated nitric acid treatment of coal bottom ash, followed by water-washing, has proved[242] to be an adequate approach given that water recovered at least 29% of the metals. Rare earths have also been extracted[243] from acid mine-drainage precipitates in passive treatment beds of the Appalachian coal basin. The 3-phase extraction process includes excavation and transportation of the precipitates, multi-phase pH-controlled step-leaching of rare earths and solvent extraction. This produces a good grade of rare-earth oxides that can be reduced to pure metal.

Slag

The distributions of indium, gallium, germanium and tin between metallic copper and lime-free or lime-containing alumina iron silicate slags and between solid Al-Fe spinel and slags have been studied[244] in situations which simulated high alumina-bearing copper scrap smelting under oxygen partial pressures of 10^{-10} to 10^{-5}atm at 1300C. This showed show that tin and indium could be efficiently taken into the copper phase under reducing conditions ($p_{O2} < 10^{-7}$atm) while gallium dissolved preferentially in the solid spinel under all conditions. Gallium dissolution into the slag and spinel occurred as $GaO_{1.5}$, while the indium in spinel was $InO_{1.5}$.

Rare-earth fluoride molten-salt electrolytic slag is a potential environmental hazard that is rich in rare earths. Rare-earth recovery from molten-salt electrolysis is of the order of 91 to 93%, with some 8% is lost as molten-salt slag. A method based upon magnetic separation, sulfuric acid leaching, HF recycling, water leaching and fluorination precipitation has been applied[245] to rare-earth recovery from electrolysis slag. Rare earths, lithium and iron-containing phases in slag were first separated magnetically. Sulfuric acid leaching was then used to produce easily soluble rare-earth sulfates. Fluoride rare-earth products were obtained by HF precipitation. The main phase in the non-magnetic fraction comprised rare earths, while the iron content was only 2.90%. Under the conditions of an acid concentration of 98.00%, a temperature of 633K, a liquid/solid ratio of 2:1, a particle size of 58 to 75μm, a stirring-rate of 300rpm and reaction-time of 3h, the transformation rates of neodymium, praseodymium and dysprosium attained more than 95.00%. The temperature had a marked influence on the sulphuric-acid leaching. With increasing temperature, the equilibrium constant of the reaction gradually increased and the stable range of NdF_3 decreased while that of Nd^{3+} increased. The activation energy for neodymium transformation was 41.57kJ/mol, indicating that the leaching was controlled by interfacial chemical reactions[246]. The reaction rate increased with decreasing particle size. Sulfation is commonly used to extract rare earths from such slag. In one study[247], sulfation completely transformed powdered samples into a sulfate mixture. Roasting and water-leaching resulted in better than 95% extraction efficiencies for rare earths and lithium while iron and aluminium remained in the residue as oxides. This led to the production of rare-earth oxalates of greater than 99% purity.

The principal waste of phosphorus production contains elements such as silicon, calcium, iron, aluminium and rare earths. These could be recovered[248] by leaching with 8M nitric acid for 2h at 75C, followed by precipitation with oxalic acid and pH-adjustment using ammonia. The recovery efficiency of $\Re_2O_3$ was 88.14%, with a purity of 99.04%. The migration behavior of rare earths during the thermal decomposition[249] of Zhijin phosphorus ore, and their extraction from phosphorus slag, is such that, during decomposition, almost all of the associated rare earths enter the slag but not the ferrophosphorus or gas phases. Amorphous calcium metasilicate and calcium fluorosilicate, major components of the slag, together with the earths exist mainly in solid solution. Some 96% of the rare earths in the slag can be dissolved in HCl solution, using an acid excess ratio 1.5 and a reaction time of 50min at 50C. Rare earths in the acid solution can be separated and recycled by using oxalic acid as a precipitator and NaOH as a pH-modifier. At a pH level of 1.7 a product having a rare-earth content of 2.1wt% is obtained, with a rare-earth recovery-rate of 88%.

Red mud

This is the colloquial term for bauxite residue; a by-product of alumina production via the Bayer process. The residue is generated at the rate of 120,000,000 to 150,000,000 tons per year and contains between 0.5 and 1.7kg of rare earths per ton. It is also a hazardous waste, amounts in total to more than 4,600,000,000 tons, occupies large areas of land and contains large amounts of sodium which dissolves easily in the sub-soil water of that land. It contains, in particular, scandium. The possible separation methods include leaching using alkaline or acid solutions, ionic liquids and biological organisms[250]. The effect of temperature, leaching-time, solid/liquid-ratio and acid concentration upon the dissolution of lanthanum, cerium, praseodymium, neodymium and scandium was studied[251] and led to a method for putting the elements into solution. It was shown that 91% of the scandium and more than 80% of other rare earths could be dissolved under optimum conditions.

Contaminated soil

Phytomanagement will become a promising solution when conventional mining methods are no longer cost-effective. Phyto-extraction permits the recovery of rare earths from soils or industrial waste. Some twenty hyperaccumulator plant species, mainly ferns such as *dicranopteris dicthotom*a, are known[252] to accumulate high concentrations of rare earths (tables 28 and 29). Plant growth-promoting rhizobacteria are able to mobilize metals and/or stimulate plant development; thus increasing the quantity of rare earth which is extracted by a plant; given its then higher phyto-extraction efficiency. Rare earths are not thought to be essential to plants but, due to their divalent charge and lesser charge density, calcium can potentially replaced by trivalent cerium, europium, gadolinium, holmium, lanthanum, neodymium, praseodymium, terbium and yttrium at calcium-binding sites in the plant molecules. The rare earths, although increasing toxic at high concentrations, can stimulate plant growth at low doses. They enter plants mainly through the roots, in free ionic form. Subsequent distribution patterns are controlled by cell-wall adsorption and the phosphate precipitation of rare earths within the roots or at the root surface. Combination with a ligand (aspartic acid, glutamic acid, citric acid, malic acid, histidine) promotes rare-earth absorption and translocation from roots to erial parts. Aspartic acid, asparagine and glutamic acid, for example, stimulate lanthanum and yttrium transport in the xylem of *phytolacca americana* while histidine increases light rare earth desorption from soil, uptake by soil solution and transport to the upper parts of *dicranopteri dichotoma*. The uptake of rare earths can also be aided by increased levels of nitrogen and potassium, and an increase in rare-earth uptake in the presence of sodium and potassium has been attributed to a direct competition between rare-earths and Ca^{2+} at uptake sites … or simply to increased plant growth.

The sequential use of reductants (dithionite-citrate-bicarbonate, hydroxylamine hydrochloride), oxidants (persulphate, hypochlorite, hydrogen peroxide), alkaline solvents (sodium hydroxide, sodium bicarbonate) and organic acids (citric, oxalic) together with the biodegradable chelating agent, [S,S]-ethylene-diamine-disuccinic-acid, has been applied to 2-stage soil washing[253]. The soil in question was contaminated with copper, zinc and lead at an e-waste recycling center. The reductants effectively extracted metals by mineral dissolution, but increased the leachability and bio-accessibility of metals due to transforming iron/manganese oxides into labile fractions. Subsequent [S,S]-ethylene-diamine-disuccinic-acid washing was required in order to reduce risk. Prior washing with oxidants was partially useful because of the limited fraction of organic matter. Prior washing with alkaline solvents was also ineffective, due to metal precipitation. Prior washing with low molecular-weight organic acids improved the extraction efficiency. When compared to hydroxylamine hydrochloride, citrates and

Materials Research Foundations **119** (2022)   https://doi.org/10.21741/9781644901793

oxalates led to lower cytotoxicity and permitted higher enzyme (dehydrogenase, acid phosphatise, urease) and nutrient (nitrogen, phosphorus) activity; thus allowing for future re-use of the treated soil.

Table 28. Total concentrations of rare earths in hyperaccumulator plants

Name	Maximum Concentration in Leaves (mg/kg)
Phytolacca icosandra	13000
Dicranopteris linearis	7000
Dicraneopteris dicthotoma	3358
Carya glabra	2300
Carya cathayensis	2296
Cary tomentosa	1350
Pronephrium simplex	1234
Pronephrium triphyllum	1027
Blechnum orientale	1022
Phytolacca Americana	1012
Stenoloma chusana	725
Woodwardia japonica	367
Athyrium yokoscence	202

Table 29. Concentrations of lanthanum in hyperaccumulator plants

Name	Maximum Concentration in Leaves (mg/kg)
Asplenium ruprechtii	40
Dryopteris erythrosora	32
Asplenium filipes	25
Asplenium trichomanes	21
Asplenium hondoense	14
Asplenium subnomale	14
Asplenium ritoense	12
Dicranopteris strigose	12
Adiantum monochlamys	11

Keyword Index

Materials Research Forum LLC
https://doi.org/10.21741/9781644901793

About the Author

Dr. Fisher has wide knowledge and experience of the fields of engineering, metallurgy and solid-state physics, beginning with work at Rolls-Royce Aero Engines on turbine-blade research, related to the Concord supersonic passenger-aircraft project, which led to a BSc degree (1971) from the University of Wales. This was followed by theoretical and experimental work on the directional solidification of eutectic alloys having the ultimate aim of developing composite turbine blades. This work led to a doctoral degree (1978) from the Swiss Federal Institute of Technology (Lausanne). He then acted for many years as an editor of various academic journals, in particular *Defect and Diffusion Forum*. In recent years he has specialized in writing monographs which introduce readers to the most rapidly developing ideas in the fields of engineering, metallurgy and solid-state physics. He is co-author of the widely-cited student textbook, *Fundamentals of Solidification*. Google Scholar credits him with 8311 citations and a lifetime h-index of 17.

References

[1] Patchett, P.J., White, W.M., Feldmann, H., Kielinczuk, S., Hofmann, A.W., Earth and Planetary Science Letters, 69[2] 1984, 365-378. https://doi.org/10.1016/0012-821X(84)90195-X

[2] Hou, Z., Liu, Y., Tian, S., Yang, Z., Xie, Y., Scientific Reports, 5, 2015, 10231. https://doi.org/10.1038/srep10231

[3] Rebello, R.Z., Lima, M.T.W.D.C., Yamane, L.H., Siman, R.R., Resources, Conservation and Recycling, 153, 2020, 104557. https://doi.org/10.1016/j.resconrec.2019.104557

[4] O'Nions, R.K., Pankhurst, R.J., Earth and Planetary Science Letters, 22[4] 1974, 328-338. https://doi.org/10.1016/0012-821X(74)90142-3

[5] Paul, D.K., Potts, P.J., Chemical Geology, 18[2] 1976, 161-167. https://doi.org/10.1016/0009-2541(76)90053-X

[6] Henderson, P., Fishlock, S.J., Laul, J.C., Cooper, T.D., Conard, R.L., Boynton, W.V., Schmitt, R.A., Earth and Planetary Science Letters, 30[1] 1976, 37-49. https://doi.org/10.1016/0012-821X(76)90006-6

[7] Blaxland, A.B., Upton, B.J., Lithos, 11[4] 1978, 291-299. https://doi.org/10.1016/0024-4937(78)90036-1

[8] Compton, P., Contributions to Mineralogy and Petrology, 66[3] 1978, 283-293. https://doi.org/10.1007/BF00373412

[9] Henderson, P., Contributions to Mineralogy and Petrology, 72[1] 1980, 81-85. https://doi.org/10.1007/BF00375570

[10] Fowler, M.B., Williams, C.T., Henderson, P., Mineralogical Magazine, 47[4] 1983, 547-553. https://doi.org/10.1180/minmag.1983.047.345.16

[11] Appel, P.W.U., Precambrian Research, 20[2-4] 1983, 243-258. https://doi.org/10.1016/0301-9268(83)90075-X

[12] Appel, P.W.U., Developments in Precambrian Geology, 7[C] 1983, 135-150.

[13] Boak, J.L., Dymek, R.F., Gromet, L.P., Gronlands Geologiske Undersogelse Rapport, 112, 1983, 23-33. https://doi.org/10.34194/rapggu.v112.7810

[14] Campbell, L.S., Henderson, P., Wall, F., Nielsen, T.F.D., Mineralogical Magazine, 61[2] 1997, 197-212. https://doi.org/10.1180/minmag.1997.061.405.04

[15] Coulson, I.M., Chambers, A.D., Canadian Mineralogist, 34[6] 1996, 1163-1178.

[16] Rae, D.A., Coulson, I.M., Chambers, A.D., Mineralogical Magazine, 60[1] 1996, 207-220. https://doi.org/10.1180/minmag.1996.060.398.14

[17] Hansen, K., Lithos, 17[C] 1984, 77-85. https://doi.org/10.1016/0024-4937(84)90007-0

[18] Appel, P.W.U., Mahabaleswar, B., Journal of the Geological Society of India, 32[3] 1998, 214-226.

[19] Whitehouse, M.J., Kamber, B.S., Precambrian Research, 126[3-4] 2003, 363-377. https://doi.org/10.1016/S0301-9268(03)00105-0

[20] Steenfelt, A., Geochemistry: Exploration, Environment, Analysis, 12[4] 2012, 313-326. https://doi.org/10.1144/geochem2011-113

[21] Friis, H., Mineralogical Magazine, 80[1] 2016, 31-41. https://doi.org/10.1180/minmag.2016.080.047

[22] Da, Z., Ma, Y., Guo, Y., Ding, Y., Zhang, W., Zhao, Y., Journal of the Chinese Rare Earth Society, 39[5] 2021, 711-722.

[23] Su, J., Gao, Y., Ni, S., Xu, R., Sun, X., Journal of Hazardous Materials, 406, 2021, 124654. https://doi.org/10.1016/j.jhazmat.2020.124654

[24] Eun, H.C., Cho, Y.Z., Son, S.M., Lee, T.K., Yang, H.C., Kim, I.T., Lee, H.S., Journal of Nuclear Materials, 420[1-3] 2012, 548-553. https://doi.org/10.1016/j.jnucmat.2011.11.048

[25] Ren, K., Tang, X., Wang, P., Willerström, J., Höök, M., Energy, 227, 2021, 120524. https://doi.org/10.1016/j.energy.2021.120524

[26] Morimoto, S., Kuroki, H., Narita, H., Ishigaki, A.. Journal of Material Cycles and Waste Management, 23[6] 2021, 2120-2132. https://doi.org/10.1007/s10163-021-01277-6

[27] Henríquez-Hernández, L.A., Romero, D., González-Antuña, A., Gonzalez-Alzaga, B., Zumbado, M., Boada, L.D., Hernández, A.F., López-Flores, I., Luzardo, O.P., Lacasaña, M., Science of the Total Environment, 706, 2020, 135750. https://doi.org/10.1016/j.scitotenv.2019.135750

[28] González-Antuña, A., Camacho, M., Henríquez-Hernández, L.A., Boada, L.D., Almeida-González, M., Zumbado, M., Luzardo, O.P., MethodsX, 4, 2017, 328-334. https://doi.org/10.1016/j.mex.2017.10.001

[29] Henríquez-Hernández, L.A., González-Antuña, A., Boada, L.D., Carranza, C., Pérez-Arellano, J.L., Almeida-González, M., Camacho, M., Zumbado, M., Fernández-Fuertes, F., Tapia-Martín, M., Luzardo, O.P., Science of the Total Environment, 636, 2018, 709-716. https://doi.org/10.1016/j.scitotenv.2018.04.311

[30] Henríquez-Hernández, L.A., Boada, L.D., Carranza, C., Pérez-Arellano, J.L., González-Antuña, A., Camacho, M., Almeida-González, M., Zumbado, M., Luzardo, O.P., Environment International, 109, 2017, 20-28. https://doi.org/10.1016/j.envint.2017.08.023

[31] Guo, C., Wei, Y., Yan, L., Li, Z., Qian, Y., Liu, H., Li, Z., Li, X., Wang, Z., Wang, J., Chemosphere, 252, 2020, 126488. https://doi.org/10.1016/j.chemosphere.2020.126488

[32] Takyi, S.A., Basu, N., Arko-Mensah, J., Dwomoh, D., Houessionon, K.G., Fobil, J.N., Chemosphere, 280, 2021, 130677. https://doi.org/10.1016/j.chemosphere.2021.130677

[33] Pagano, G., Thomas, P.J., Di Nunzio, A., Trifuoggi, M., Environmental Research, 171, 2019, 493-500. https://doi.org/10.1016/j.envres.2019.02.004

[34] Zumbado, M., Luzardo, O.P., Rodríguez-Hernández, Á., Boada, L.D., Henríquez-Hernández, L.A., Environmental Research, 169, 2019, 368-376. https://doi.org/10.1016/j.envres.2018.11.021

[35] Gwenzi, W., Mangori, L., Danha, C., Chaukura, N., Dunjana, N., Sanganyado, E., Science of the Total Environment, 636, 2018, 299-313. https://doi.org/10.1016/j.scitotenv.2018.04.235

[36] Freitas, R., Cardoso, C.E.D., Costa, S., Morais, T., Moleiro, P., Lima, A.F.D., Soares, M., Figueiredo, S., Águeda, T.L., Rocha, P., Amador, G., Soares, A.M.V.M., Pereira, E., Environmental Pollution, 260, 2020, 113859. https://doi.org/10.1016/j.envpol.2019.113859

[37] Pinto, J., Costa, M., Leite, C., Borges, C., Coppola, F., Henriques, B., Monteiro, R., Russo, T., Di Cosmo, A., Soares, A.M.V.M., Polese, G., Pereira, E., Freitas, R., Aquatic Toxicology, 211, 2019, 181-192. https://doi.org/10.1016/j.aquatox.2019.03.017

[38] Moreira, A., Henriques, B., Leite, C., Libralato, G., Pereira, E., Freitas, R., Ecological Indicators, 108, 2020, 105687. https://doi.org/10.1016/j.ecolind.2019.105687

[39] Fujita, Y., Barnes, J., Eslamimanesh, A., Lencka, M.M., Anderko, A., Riman, R.E., Navrotsky, A., Environmental Science and Technology, 49[16] 2015, 9460-9468. https://doi.org/10.1021/acs.est.5b01753

[40] Rasoulnia, P., Barthen, R., Puhakka, J.A., Lakaniemi, A.M., Journal of Hazardous Materials, 414, 2021, 125564. https://doi.org/10.1016/j.jhazmat.2021.125564

[41] Rasoulnia, P., Barthen, R., Valtonen, K., Lakaniemi, A.M., Waste and Biomass Valorization, 12[10] 2021, 5545-5559. https://doi.org/10.1007/s12649-021-01398-x

[42] Anshu, P., Hait, S., Waste Management, 75, 2018, 103-123. https://doi.org/10.1016/j.wasman.2018.02.014

[43] Kucuker, M.A., Kuchta, K., Global Nest Journal, 20[4] 2018) 737-742. https://doi.org/10.30955/gnj.002692

[44] Auerbach, R., Bokelmann, K., Stauber, R., Gutfleisch, O., Schnell, S., Ratering, S., Minerals Engineering, 134, 2019, 104-117. https://doi.org/10.1016/j.mineng.2018.12.022

[45] Hopfe, S., Konsulke, S., Barthen, R., Lehmann, F., Kutschke, S., Pollmann, K., Waste Management, 79, 2018, 554-563. https://doi.org/10.1016/j.wasman.2018.08.030

[46] Marra, A., Cesaro, A., Rene, E.R., Belgiorno, V., Lens, P.N.L., Journal of Environmental Management, 210, 2018, 180-190. https://doi.org/10.1016/j.jenvman.2017.12.066

[47] Monneron-Enaud, B., Wiche, O., Schlömann, M., Recycling, 5[3] 2020, 1-14. https://doi.org/10.3390/recycling5030022

[48] Di Piazza, S., Cecchi, G., Cardinale, A.M., Carbone, C., Mariotti, M.G., Giovine, M., Zotti, M., Waste Management, 60, 2017, 596-600. https://doi.org/10.1016/j.wasman.2016.07.029

[49] Yang, H., Yan, C., Luo, W., Liu, C., Zhou, Q., Journal of the Taiwan Institute of Chemical Engineers, 74, 2017, 105-112. https://doi.org/10.1016/j.jtice.2017.02.002

[50] Bai, R., Yang, F., Zhang, Y., Zhao, Z., Liao, Q., Chen, P., Zhao, P., Guo, W., Cai, C., Carbohydrate Polymers, 190, 2018, 255-261. https://doi.org/10.1016/j.carbpol.2018.02.059

[51] Brewer, A., Dohnalkova, A., Shutthanandan, V., Kovarik, L., Chang, E., Sawvel, A.M., Mason, H.E., Reed, D., Ye, C., Hynes, W.F., Lammers, L.N., Park, D.M., Jiao, Y., Environmental Science and Technology, 53[23] 2019, 13888-13897. https://doi.org/10.1021/acs.est.9b04608

[52] Wang, Q., Wilfong, W.C., Kail, B.W., Yu, Y., Gray, M.L., ACS Sustainable Chemistry and Engineering, 5[11] 2017, 10947-10958. https://doi.org/10.1021/acssuschemeng.7b02851

[53] Deblonde, G.J.P., Mattocks, J.A., Park, D.M., Reed, D.W., Cotruvo, J.A., Jiao, Y., Inorganic Chemistry, 59[17] 2020, 11855-11867. https://doi.org/10.1021/acs.inorgchem.0c01303

[54] Firkala, T., Lederer, F., Pollmann, K., Rudolph, M., Solid State Phenomena, 262, 2017, 537-540. https://doi.org/10.4028/www.scientific.net/SSP.262.537

[55] Lederer, F.L., Curtis, S.B., Bachmann, S., Dunbar, W.S., MacGillivray, R.T.A., Biotechnology and Bioengineering, 114[5] 2017, 1016-1024. https://doi.org/10.1002/bit.26240

[56] Shu, Q., Liao, C.F., Zou, W.Q., Xu, B.Q., Tan, Y.H., Transactions of Nonferrous Metals Society of China, 31[4] 2021, 1127-1139. https://doi.org/10.1016/S1003-6326(21)65566-8

[57] Mezy, A., Vardanyan, A., Garcia, A., Schmitt, C., Lakić, M., Krajnc, S., Daniel, G., Košak, A., Lobnik, A., Seisenbaeva, G.A., Separation and Purification Technology, 276, 2021, 119340. https://doi.org/10.1016/j.seppur.2021.119340

[58] Čížková, M., Mezricky, P., Mezricky, D., Rucki, M., Zachleder, V., Vítová, M., Waste and Biomass Valorization, 12[6] 2021, 3137-3146. https://doi.org/10.1007/s12649-020-01182-3

[59] Goecke, F., Zachleder, V., Vítová, M., Algal Biorefineries, Volume 2, 2015, 339-363. https://doi.org/10.1007/978-3-319-20200-6_10

[60] Cao, Y., Shao, P., Chen, Y., Zhou, X., Yang, L., Shi, H., Yu, K., Luo, X., Luo, X., Resources, Conservation and Recycling, 169, 2021, 105519. https://doi.org/10.1016/j.resconrec.2021.105519

[61] Pinto, J., Costa, M., Henriques, B., Soares, J., Dias, M., Viana, T., Ferreira, N., Vale, C., Pinheiro-Torres, J., Pereira, E., Journal of Rare Earths, 39[6] 2021, 734-741. https://doi.org/10.1016/j.jre.2020.09.025

[62] Pinto, J., Costa, M., Henriques, B., Soares, J., Dias, M., Viana, T., Ferreira, N., Vale, C., Pinheiro-Torres, J., Pereira, E., Journal of Rare Earths. 39[6] 2021, 734-741. https://doi.org/10.1016/j.jre.2020.09.025

[63] Henriques, B., Morais, T., Cardoso, C.E.D., Freitas, R., Viana, T., Ferreira, N., Fabre, E., Pinheiro-Torres, J., Pereira, E., Science of the Total Environment, 786, 2021, 147176. https://doi.org/10.1016/j.scitotenv.2021.147176

[64] Fabre, E., Henriques, B., Viana, T., Pinto, J., Costa, M., Ferreira, N., Tavares, D., Vale, C., Pinheiro-Torres, J., Pereira, E., Journal of Environmental Chemical Engineering, 9[5] 2021, 105946. https://doi.org/10.1016/j.jece.2021.105946

[65] Hamza, M.F., Salih, K.A.M., Abdel-Rahman, A.A.H., Zayed, Y.E., Wei, Y., Liang, J., Guibal, E., Chemical Engineering Journal, 403, 2021, 126399. https://doi.org/10.1016/j.cej.2020.126399

[66] Wei, Y., Salih, K.A.M., Rabie, K., Elwakeel, K.Z., Zayed, Y.E., Hamza, M.F., Guibal, E., Chemical Engineering Journal, 412, 2021, 127399. https://doi.org/10.1016/j.cej.2020.127399

[67] Wei, Y., Salih, K.A.M., Hamza, M.F., Rodríguez Castellón, E., Guibal, E., Chemical Engineering Journal, 424, 2021, 130500. https://doi.org/10.1016/j.cej.2021.130500

[68] Wei, Y., Salih, K.A.M., Hamza, M.F., Fujita, T., Rodríguez-Castellón, E., Guibal, E., Polymers, 13[9] 2021, 1513. https://doi.org/10.3390/polym13091513

[69] He, C., Salih, K.A.M., Wei, Y., Mira, H., Abdel-Rahman, A.A.H., Elwakeel, K.Z., Hamza, M.F., Guibal, E., Metals, 11[2] 2021, 1-29. https://doi.org/10.3390/met11020294

[70] Hopfe, S., Flemming, K., Lehmann, F., Möckel, R., Kutschke, S., Pollmann, K., Waste Management, 62, 2017, 211-221. https://doi.org/10.1016/j.wasman.2017.02.005

[71] Gupta, N.K., Choudhary, B.C., Gupta, A., Achary, S.N., Sengupta, A., Journal of Molecular Liquids, 289, 2019, 111121. https://doi.org/10.1016/j.molliq.2019.111121

[72] Cole, B.E., Falcones, I.B., Cheisson, T., Manor, B.C., Carroll, P.J., Schelter, E.J., Chemical Communications, 54[73] 2018, 10276-10279. https://doi.org/10.1039/C8CC04409K

[73] Peng, X., Mo, S., Li, R., Li, J., Tian, C., Liu, W., Wang, Y., Environmental Chemistry Letters, 19[1] 2021, 719-728. https://doi.org/10.1007/s10311-020-01073-y

[74] Zhou, J., Song, X., Shui, B., Wang, S., Coatings, 11[9] 2021, 1040. https://doi.org/10.3390/coatings11091040

[75] Liu, Z., Chen, G., Li, X., Lu, X., Journal of Alloys and Compounds, 856, 2021, 158185. https://doi.org/10.1016/j.jallcom.2020.158185

[76] Li, X., Lan, J., Hu, W., Yang, Y., Li, Y., Ion Exchange and Adsorption, 30[6] 2014, 499-506.

[77] Ali, I., Zakharchenko, E.A., Myasoedova, G.V., Molochnikova, N.P., Rodionova, A.A., Baulin, V.E., Burakov, A.E., Burakova, I.V., Babkin, A.V., Neskoromnaya, E.A., Melezhik, A.V., Tkachev, A.G., Habila, M.A., El-Marghany, A., Sheikh, M., Ghfar, A., Journal of Molecular Liquids, 335, 2021, 116260. https://doi.org/10.1016/j.molliq.2021.116260

[78] O'Connor, M.P., Coulthard, R.M., Plata, D.L., Environmental Science: Water Research and Technology, 4[1] 2018, 58-66. https://doi.org/10.1039/C7EW00187H

[79] Babu, C.M., Binnemans, K., Roosen, J., Industrial and Engineering Chemistry Research, 57[5] 2018, 1487-1497. https://doi.org/10.1021/acs.iecr.7b04274

[80] Zubiani, E.M.I., Cristiani, C., Dotelli, G., Stampino, P.G., International Conference on Wastes, 2015, 361-366. https://doi.org/10.1201/b18853-60

[81] Patra, S., Roy, E., Madhuri, R., Sharma, P.K., ACS Sustainable Chemistry and Engineering, 5[8] 2017, 6910-6923. https://doi.org/10.1021/acssuschemeng.7b01124

[82] Artiushenko, O., Zaitsev, V., Rojano, W.S., Freitas, G.A., Nazarkovsky, M., Saint'Pierre, T.D., Kai, J., Journal of Hazardous Materials, 408, 2021, 124976. https://doi.org/10.1016/j.jhazmat.2020.124976

[83] Seisenbaeva, G.A., Ali, L.M.A., Vardanyan, A., Gary-Bobo, M., Budnyak, T.M., Kessler, V.G., Durand, J.O., Journal of Hazardous Materials, 406, 2021, 124698. https://doi.org/10.1016/j.jhazmat.2020.124698

[84] Salih, K.A.M., Hamza, M.F., Mira, H., Wei, Y., Gao, F., Atta, A.M., Fujita, T., Guibal, E., Molecules, 26[4] 2021, 1049. https://doi.org/10.3390/molecules26041049

[85] Kobylinska, N., Mishra, B., Kessler, V.G., Tripathi, B.P., Seisenbaeva, G.A., Microporous and Mesoporous Materials, 315, 2021, 110919. https://doi.org/10.1016/j.micromeso.2021.110919

[86] Polyakov, E.G., Sibilev, A.S., Metallurgist, 59[5-6] 2015, 368-373. https://doi.org/10.1007/s11015-015-0111-8

[87] Le, T.H., Malfliet, A., Blanpain, B., Guo, M., REWAS: Towards Materials Resource Sustainability, 2016, 95-100. https://doi.org/10.1002/9781119275039.ch14

[88] Sahle-Demessie, E., Mezgebe, B., Dietrich, J., Shan, Y., Harmon, S., Lee, C.C., Journal of Environmental Chemical Engineering, 9[1] 2021, 104943. https://doi.org/10.1016/j.jece.2020.104943

[89] Diaz, L.A., Lister, T.E., Parkman, J.A., Clark, G.G., Journal of Cleaner Production, 125, 2016, 236-244. https://doi.org/10.1016/j.jclepro.2016.03.061

[90] Ishii, M., Matsumiya, M., Kawakami, S., ECS Transactions, 50[11] 2012, 549-560. https://doi.org/10.1149/05011.0549ecst

[91] Shimohara, Y., Nezu, A., Numakura, M., Akatsuka, H., Matsuura, H., Molten Salts Chemistry and Technology, 9781118448731, 2014, 577-580. https://doi.org/10.1002/9781118448847.ch7g

[92] Cvetković, V.S., Feldhaus, D., Vukićević, N.M., Barudžija, T.S., Friedrich, B., Jovićević, J.N., Metals, 11[9] 2021, 1494. https://doi.org/10.3390/met11091494

[93] Zhang, B., Wang, L., Liu, Y., Zhang, Y., Zhang, L., Shi, Z., Separation and Purification Technology, 276, 2021, 119416. https://doi.org/10.1016/j.seppur.2021.119416

[94] Ishioka, K., Matsumiya, M., Ishii, M., Kawakami, S., Hydrometallurgy, 144-145, 2014, 186-194. https://doi.org/10.1016/j.hydromet.2014.02.007

[95] Lyman, J.W., Palmer, G.R., High Temperature Materials and Processes, 11[1-4] 1993, 175-188. https://doi.org/10.1515/HTMP.1993.11.1-4.175

[96] Zhang, J., Azimi, G., Metals and Materials Series, 2020, 93-105. https://doi.org/10.1007/978-3-030-36758-9_9

[97] Ippolito, N.M., Ferella, F., Innocenzi, V., Trapasso, F., Passeri, D., Belardi, G., Vegliò, F., Minerals Engineering, 167, 2021, 106906. https://doi.org/10.1016/j.mineng.2021.106906

[98] Zhang, C., Tang, D., Cao, M., Gu, F., Cai, X., Liu, X., Cheng, Z., Hall, P., Fu, J., Zhao, P., Resources, Conservation and Recycling, 174, 2021, 105769. https://doi.org/10.1016/j.resconrec.2021.105769

[99] Liu, H., Li, S., Wang, B., Wang, K., Wu, R., Ekberg, C., Volinsky, A.A., Journal of Cleaner Production, 238, 2019, 117998. https://doi.org/10.1016/j.jclepro.2019.117998

[100] Tanvar, H., Shukla, N., Dhawan, N., Journal of Metals, 72[2] 2020, 823-830. https://doi.org/10.1007/s11837-019-03890-1

[101] Geng, A., Zhu, Z., Hua, Z., Chen, J., Wu, B., He, S., Chinese Journal of Rare Metals, 45[4] 2021, 455-464.

[102] Kumari, A., Raj, R., Randhawa, N.S., Sahu, S.K., Hydrometallurgy, 201, 2021, 105581. https://doi.org/10.1016/j.hydromet.2021.105581

[103] Lim, K.H., Choi, C.U., Moon, G., Lee, T.H., Kang, J., Journal of Sustainable Metallurgy, 7[3] 2021, 794-805. https://doi.org/10.1007/s40831-021-00380-0

[104] Liu, F., Chen, F., Wang, L., Ma, S., Wan, X., Wang, J., Hydrometallurgy, 203, 2021, 105626. https://doi.org/10.1016/j.hydromet.2021.105626

[105] Liu, X., Huang, L., Liu, Z., Zhang, D., Gao, K., Li, M., Journal of Cleaner Production, 321, 2021, 128784. https://doi.org/10.1016/j.jclepro.2021.128784

[106] Han, K.N., Kim, R., Minerals, 11[7] 2021, 670. https://doi.org/10.3390/min11070670

[107] Virtanen, E.J., Perämäki, S., Helttunen, K., Väisänen, A., Moilanen, J.O., ACS Omega, 6[37] 2021, 23977-23987. https://doi.org/10.1021/acsomega.1c02982

[108] Yu, J.F., Fu, J., Cheng, H., Cui, Z., Waste Management, 61, 2017, 362-371. https://doi.org/10.1016/j.wasman.2016.12.014

[109] Deshmane, V.G., Islam, S.Z., Bhave, R.R., Environmental Science and Technology. 54[1] 2020, 550-558. https://doi.org/10.1021/acs.est.9b05695

[110] Tanaka, M., Narita, H., Hydrometallurgy, 201, 2021, 105588. https://doi.org/10.1016/j.hydromet.2021.105588

[111] Williams-Wynn, M.D., Naidoo, P., Ramjugernath, D., Minerals Engineering, 153, 2020, 106285. https://doi.org/10.1016/j.mineng.2020.106285

[112] Schaeffer, N., Passos, H., Billard, I., Papaiconomou, N., Coutinho, J.A.P., Critical Reviews in Environmental Science and Technology, 48[13-15] 2018, 859-922. https://doi.org/10.1080/10643389.2018.1477417

[113] Hu, A.H., Kuo, C.H., Huang, L.H., Su, C.C., Waste Management, 60, 2017, 765-774. https://doi.org/10.1016/j.wasman.2016.10.032

[114] Zhang, Y., Guo, W., Liu, D., Xu, J., Journal of Rare Earths, 39[11] 2021, 1435-1441. https://doi.org/10.1016/j.jre.2020.10.005

[115] Orefice, M., Binnemans, K., Separation and Purification Technology, 258, 2021, 117800. https://doi.org/10.1016/j.seppur.2020.117800

[116] Cui, H., Shi, J., Liu, Y., Yan, N., Zhang, C., You, S., Chen, G., Separation and Purification Technology, 273, 2021, 119010. https://doi.org/10.1016/j.seppur.2021.119010

[117] Joo, S.H., Shin, D., Oh, C., Wang, J.P., Shin, S.M., 14th International Symposium on East Asian Resources Recycling Technology, 2017.

[118] Rinne, M., Elomaa, H., Porvali, A., Lundström, M., Resources, Conservation and Recycling, 170, 2021, 105586. https://doi.org/10.1016/j.resconrec.2021.105586

[119] Meshram, P., Pandey, B.D., Mankhand, T.R., Waste Management, 51, 2016, 196-203. https://doi.org/10.1016/j.wasman.2015.12.018

[120] Lin, S.L., Huang, K.L., Wang, I.C., Chou, I.C., Kuo, Y.M., Hung, C.H., Lin, C., Journal of the Air and Waste Management Association, 66[3] 2016, 296-306. https://doi.org/10.1080/10962247.2015.1131206

[121] Ahn, N.K., Shim, H.W., Kim, D.W., Swain, B., Waste Management, 104, 2020, 254-261. https://doi.org/10.1016/j.wasman.2020.01.014

[122] De Oliveira Dos Santos, V.E., Celante, V.G., De Fátima Fontes Lelis, M., De Freitas, M.B.J.G., Quimica Nova, 37[1] 2014, 22-26. https://doi.org/10.1590/S0100-40422014000100005

[123] Alonso, A.R., Pérez, E.A., Lapidus, G.T., Luna-Sánchez, R.M., Canadian Metallurgical Quarterly, 54[3] 2015, 310-317. https://doi.org/10.1179/1879139515Y.0000000013

[124] de Oliveira, W.C.M., Rodrigues, G.D., Mageste, A.B., de Lemos, L.R., Chemical Engineering Journal, 322, 2017, 346-352. https://doi.org/10.1016/j.cej.2017.04.044

[125] Yao, Y., Farac, N., Azimi, G., ECS Transactions, 85[13] 2018, 405-415. https://doi.org/10.1149/08513.0405ecst

[126] Yao, Y., Farac, N.F., Azimi, G., ACS Sustainable Chemistry and Engineering, 6[1] 2018, 1417-1426. https://doi.org/10.1021/acssuschemeng.7b03803

[127] Porvali, A., Ojanen, S., Wilson, B.P., Serna-Guerrero, R., Lundström, M., Journal of Sustainable Metallurgy, 6[1] 2020, 78-90. https://doi.org/10.1007/s40831-019-00258-2

[128] Su, X., Xie, W., Sun, X., Minerals Engineering, 160, 2021, 106641. https://doi.org/10.1016/j.mineng.2020.106641

[129] Maroufi, S., Nekouei, R.K., Hossain, R., Assefi, M., Sahajwalla, V., ACS Sustainable Chemistry and Engineering, 6[9] 2018, 11811-11818. https://doi.org/10.1021/acssuschemeng.8b02097

[130] Zielinski, M., Cassayre, L., Destrac, P., Coppey, N., Garin, G., Biscans, B., ChemSusChem, 13[3] 2020, 616-628. https://doi.org/10.1002/cssc.201902640

[131] Zielinski, M., Cassayre, L., Coppey, N., Biscans, B., Crystal Growth and Design, 21[10] 2021, 5943-5954. https://doi.org/10.1021/acs.cgd.1c00834

[132] Porvali, A., Agarwal, V., Lundström, M., Waste Management, 107, 2020, 66-73. https://doi.org/10.1016/j.wasman.2020.03.042

[133] Porvali, A., Wilson, B.P., Lundström, M., Waste Management, 71, 2018, 381-389. https://doi.org/10.1016/j.wasman.2017.10.031

[134] Zhang, Y., Chu, W., Chen, X., Wang, M., Cui, H., Wang, J., Journal of Cleaner Production, 235, 2019, 1295-1303. https://doi.org/10.1016/j.jclepro.2019.07.072

[135] Tang, K., Ciftja, A., Van Der Eijk, C., Wilson, S., Tranell, G., Journal of Mining and Metallurgy, Section B: Metallurgy, 49[2] 2013, 233-236. https://doi.org/10.2298/JMMB120808004T

[136] Yang, X., Zhang, J., Fang, X., Journal of Hazardous Materials, 279, 2014, 384-388. https://doi.org/10.1016/j.jhazmat.2014.07.027

[137] He, H.W., Meng, J., Journal of Central South University - Science and Technology, 42[9] 2011, 2651-2657.

[138] Ilyas, S., Kim, H., Srivastava, R.R., Journal of Metals, 73[1] 2021, 19-26. https://doi.org/10.1007/s11837-020-04471-3

[139] Abo Atia, T., Wouters, W., Monforte, G., Spooren, J., Resources, Conservation and Recycling, 166, 2021, 105349. https://doi.org/10.1016/j.resconrec.2020.105349

[140] Sposato, C., Catizzone, E., Blasi, A., Forte, M., Romanelli, A., Morgana, M., Braccio, G., Giordano, G., Migliori, M., Processes, 9[8] 2021, 1369. https://doi.org/10.3390/pr9081369

[141] Maidel, M., Ponte, M.J.J.D.S., Ponte, H.D.A., Valt, R.B.G., Separation and Purification Technology, 281, 2022, 119905. https://doi.org/10.1016/j.seppur.2021.119905

[142] Yin, X., Yu, J., Wu, Y., Tian, X., Wang, W., Zhang, Y.N., Zuo, T., ACS Sustainable Chemistry and Engineering, 6[3] 2018, 4321-4329. https://doi.org/10.1021/acssuschemeng.7b04796

[143] De La Torre, E., Vargas, E., Ron, C., Gámez, S., Metals, 8[10] 2018, 777. https://doi.org/10.3390/met8100777

[144] Lie, J., Ismadji, S., Liu, J.C., Journal of Chemical Technology and Biotechnology, 94[12] 2019, 3859-3865. https://doi.org/10.1002/jctb.6184

[145] Pindar, S., Dhawan, N., Resources, Conservation and Recycling, 169, 2021, 105469. https://doi.org/10.1016/j.resconrec.2021.105469

[146] Toache-Pérez, A.D., Bolarín-Miró, A.M., Sánchez-De Jesús, F., Lapidus, G.T., Sustainable Environment Research, 30[1] 2020, 20. https://doi.org/10.1186/s42834-020-00060-w

[147] Otsuki, A., Dodbiba, G., Fujita, T., 20th International Microprocesses and Nanotechnology Conference, MNC, 4456191, 2007, 236-237.

[148] De Michelis, I., Ferella, F., Varelli, E.F., Vegliò, F., Waste Management, 31[12] 2011, 2559-2568. https://doi.org/10.1016/j.wasman.2011.07.004

[149] Saratale, G.D., Kim, H.Y., Saratale, R.G., Kim, D.S., Minerals Engineering, 152, 2020, 106341. https://doi.org/10.1016/j.mineng.2020.106341

[150] Song, G., Yuan, W., Zhu, X., Wang, X., Zhang, C., Li, J., Bai, J., Wang, J., Journal of Cleaner Production, 151, 2017, 361-370. https://doi.org/10.1016/j.jclepro.2017.03.086

[151] Shukla, N., Tanvar, H., Dhawan, N., Physicochemical Problems of Mineral Processing, 56[4] 2020, 710-722.

[152] Esbrí, J.M., Rivera, S., Tejero, J., Higueras, P.L., Environmental Science and Pollution Research, 28[43] 2021, 61860-61868. https://doi.org/10.1007/s11356-021-16800-3

[153] Pavón, S., Lapo, B., Fortuny, A., Sastre, A.M., Bertau, M., Separation and Purification Technology, 272, 2021, 118879. https://doi.org/10.1016/j.seppur.2021.118879

[154] Pavón, S., Lorenz, T., Fortuny, A., Sastre, A.M., Bertau, M., Waste Management, 122, 2021, 55-63. https://doi.org/10.1016/j.wasman.2020.12.039

[155] Lorenz, T., Golon, K., Fröhlich, P., Bertau, M., Chemie-Ingenieur-Technik, 87[10] 2015, 1373-1382. https://doi.org/10.1002/cite.201400181

[156] Rodriguez, N.R., Grymonprez, B., Binnemans, K., Industrial and Engineering Chemistry Research, 60[28] 2021, 10319-10326. https://doi.org/10.1021/acs.iecr.1c01429

[157] Tunsu, C., Ekberg, C., Foreman, M., Retegan, T., Transactions of the Institutions of Mining and Metallurgy C, 125[4] 2016, 199-203. https://doi.org/10.1080/03719553.2016.1181398

[158] Kochmanova, A., Miskufova, A., Palencar, M., Horvathova, H., Metall, 70[5] 2016, 185-189.

[159] Van Loy, S., Binnemans, K., Van Gerven, T., Journal of Cleaner Production, 156, 2017, 226-234. https://doi.org/10.1016/j.jclepro.2017.03.160

[160] Dupont, D., Binnemans, K., Green Chemistry, 17[2] 2015, 856-868. https://doi.org/10.1039/C4GC02107J

[161] Bengio, D., Dumas, T., Arpigny, S., Husar, R., Mendes, E., Solari, P.L., Schlegel, M.L., Schlegel, D., Pellet-Rostaing, S., Moisy, P., Chemistry - a European Journal, 26[63] 2020, 14385-14396. https://doi.org/10.1002/chem.202001469

[162] Otsuki, A., Dodbiba, G., Shibayama, A., Sadaki, J., Mei, G., Fujita, T., Proceedings - European Metallurgical Conference, EMC, 3, 2007, 1507-1519.

[163] Patil, A.B., Tarik, M., Struis, R.P.W.J., Ludwig, C., Resources, Conservation and Recycling, 164, 2021, 105153. https://doi.org/10.1016/j.resconrec.2020.105153

[164] Liu, H., Zhang, S., Pan, D., Tian, J., Yang, M., Wu, M., Volinsky, A.A., Journal of Hazardous Materials, 272, 2014, 96-101. https://doi.org/10.1016/j.jhazmat.2014.02.043

[165] Pavón, S., Fortuny, A., Coll, M.T., Sastre, A.M., Waste Management, 82, 2018, 241-248. https://doi.org/10.1016/j.wasman.2018.10.027

[166] He, L., Ji, W., Yin, Y., Sun, W., Journal of Rare Earths, 36[1] 2018, 108-112. https://doi.org/10.1016/j.jre.2017.05.016

[167] Ricci, P.C., Murgia, M., Carbonaro, C.M., Sgariotto, S., Stagi, L., Corpino, R., Chiriu, D., Grilli, M.L., IOP Conference Series - Materials Science and Engineering, 329[1] 2018, 012002. https://doi.org/10.1088/1757-899X/329/1/012002

[168] Ott, B., Spiller, D.E., Taylor, P.R., Minerals, Metals and Materials Series, 2019, 295-304. https://doi.org/10.1007/978-3-030-10386-6_34

[169] München, D.D., Veit, H.M., Waste Management, 61, 2017, 372-376. https://doi.org/10.1016/j.wasman.2017.01.032

[170] Tanvar, H., Kumar, S., Dhawan, N., Journal of Metals, 71[7] 2019, 2345-2352. https://doi.org/10.1007/s11837-019-03523-7

[171] Yang, Y., Lan, C., Wang, Y., Zhao, Z., Li, B., Separation and Purification Technology, 230, 2020, 115870. https://doi.org/10.1016/j.seppur.2019.115870

[172] Tian, Y., Zhou, X., Yu, H., Liu, Z., Zhan, H., Chinese Journal of Rare Metals, 44[4] 2020, 427-432.

[173] Asabe, K., Saguchi, A., Takahashi, W., Suzuki, R.O., Ono, K., Materials Transactions, 42[12] 2001, 2487-2491. https://doi.org/10.2320/matertrans.42.2487

[174] Saguchi, A., Asabe, K., Fukuda, T., Takahashi, W., Suzuki, R.O., Journal of Alloys and Compounds, 408-412, 2006, 1377-1381. https://doi.org/10.1016/j.jallcom.2005.04.178

[175] Suzuki, R.O., Saguchi, A., Takahashi, W., Yagura, T., Ono, K., Materials Transactions, 42[12] 2001, 2492-2498. https://doi.org/10.2320/matertrans.42.2492

[176] Saguchi, A., Asabe, K., Takahashi, W., Suzuki, R.O., Ono, K., Materials Transactions, 43[2] 2002, 256-260. https://doi.org/10.2320/matertrans.43.256

[177] Maroufi, S., Khayyam Nekouei, R., Sahajwalla, V.. ACS Sustainable Chemistry and Engineering, 5[7] 2017, 6201-6208. https://doi.org/10.1021/acssuschemeng.7b01133

[178] Lalana, E.H., Degri, M.J.J., Bradshaw, A., Walton, A., World PM Congress and Exhibition, 2016.

[179] Li, C., Liu, W.Q., Yue, M., Liu, Y.Q., Zhang, D.T., Zuo, T.Y., IEEE Transactions on Magnetics, 50[12] 2014, 6832570. https://doi.org/10.1109/TMAG.2014.2329457

[180] Chinnasamy, C., Jasinski, M.M., Dent, P., Liu, J., Advances in Powder Metallurgy and Particulate Materials, Proceedings of the International Conference on Powder Metallurgy and Particulate Materials, 2013, 7139-7147.

[181] Ni, S., Su, J., Zhang, H., Zeng, Z., Zhi, H., Sun, X., Chemical Engineering Journal, 412, 2021, 128602. https://doi.org/10.1016/j.cej.2021.128602

[182] Sato, N., Wei, Y., Nanjo, M., Tokuda, M., Metallurgical Review of the Mining and Metallurgical Institute of Japan, 15[1] 1998, 1-13.

[183] Itoh, M., Masuda, M., Suzuki, S., Machida, K.I., Journal of Alloys and Compounds, 374[1-2] 2004, 393-396. https://doi.org/10.1016/j.jallcom.2003.11.030

[184] Machida, K.I., Itoh, M., Masuda, M., Kojima, S., Journal of the Japan Society of Powder and Powder Metallurgy, 51[3] 2004, 160-164. https://doi.org/10.2497/jjspm.51.160

[185] Gandha, K., Ouyang, G., Gupta, S., Kunc, V., Paranthaman, M.P., Nlebedim, I.C., Waste Management, 90, 2019, 94-99. https://doi.org/10.1016/j.wasman.2019.04.040

[186] Maroufi, S., Assefi, M., Sahajwalla, V., Resources, Conservation and Recycling, 139, 2018, 172-177. https://doi.org/10.1016/j.resconrec.2018.08.014

[187] Maroufi, S., Nekouei, R.K., Sahajwalla, V., ACS Sustainable Chemistry and Engineering, 6[3] 2018, 3402-3410. https://doi.org/10.1021/acssuschemeng.7b03585

[188] Peelman, S., Sietsma, J., Yang, Y., Journal of Sustainable Metallurgy, 4[2] 2018, 276-287. https://doi.org/10.1007/s40831-018-0165-5

[189] Prodius, D., Gandha, K., Mudring, A.V., Nlebedim, I.C., ACS Sustainable Chemistry and Engineering, 8[3] 2020, 1455-1463. https://doi.org/10.1021/acssuschemeng.9b05741

[190] Cheraitia, K., Lounis, A., Mehenni, M., Separation Science and Technology, 53[1] 2018, 161-169. https://doi.org/10.1080/01496395.2017.1375956

[191] Wang, L.P., Chen, W.S., Murata, K., Dodbiba, G., Fujita, T., Proceedings of the 13th International Symposium on East Asian Resources Recycling Technology, 2015, 904-907.

[192] Iannicelli-Zubiani, E.M., Cristiani, C., Dotelli, G., Gallo Stampino, P., Pelosato, R., Mesto, E., Schingaro, E., Lacalamita, M., Waste Management, 46, 2015, 546-556. https://doi.org/10.1016/j.wasman.2015.09.017

[193] Cristiani, C., Bellotto, M., Dotelli, G., Latorrata, S., Ramis, G., Stampino, P.G., Zubiani, E.M.I., Finocchio, E., Minerals, 11[1] 2021, 1-15. https://doi.org/10.3390/min11010015

[194] Karal, E., Kucuker, M.A., Demirel, B., Copty, N.K., Kuchta, K., Journal of Cleaner Production, 288, 2020, 125087. https://doi.org/10.1016/j.jclepro.2020.125087

[195] Diehl, O., Schönfeldt, M., Brouwer, E., Dirks, A., Rachut, K., Gassmann, J., Güth, K., Buckow, A., Gauss, R., Stauber, R., Gutfleisch, O., Journal of Sustainable Metallurgy, 4[2] 2018, 163-175. https://doi.org/10.1007/s40831-018-0171-7

[196] Erust, C., Akcil, A., Tuncuk, A., Deveci, H., Yazici, E.Y., Mineral Processing and Extractive Metallurgy Review, 42[2] 2021, 90-101. https://doi.org/10.1080/08827508.2019.1692010

[197] Li, X., Li, Z., Orefice, M., Binnemans, K., ACS Sustainable Chemistry and Engineering, 7[2] 2019, 2578-2584. https://doi.org/10.1021/acssuschemeng.8b05604

[198] Li, X., Li, Z., Binnemans, K., Separation and Purification Technology, 275, 2021, 119158. https://doi.org/10.1016/j.seppur.2021.119158

[199] Kaplan, V., Wachtel, E., Gartsman, K., Feldman, Y., Park, K.T., Lubomirsky, I., Journal of Metals, 73[6] 2021, 1957-1965. https://doi.org/10.1007/s11837-021-04592-3

[200] Makarova, I., Ryl, J., Sun, Z., Kurilo, I., Górnicka, K., Laatikainen, M., Repo, E., Separation and Purification Technology, 251, 2020, 117362. https://doi.org/10.1016/j.seppur.2020.117362

[201] Dwadasi, B.S., Gupta, S., Daware, S., Srinivasan, S.G., Rai, B., Industrial and Engineering Chemistry Research, 57[50] 2018, 17209-17217. https://doi.org/10.1021/acs.iecr.8b03423

[202] Zhang, J., Anawati, J., Yao, Y., Azimi, G., ACS Sustainable Chemistry and Engineering, 6[12] 2018, 16713-16725. https://doi.org/10.1021/acssuschemeng.8b03992

[203] Dupont, D., Binnemans, K., Green Chemistry, 17[4] 2015, 2150-2163. https://doi.org/10.1039/C5GC00155B

[204] Vander Hoogerstraete, T., Blanpain, B., Van Gerven, T., Binnemans, K., RSC Advances, 4[109] 2014, 64099-64111. https://doi.org/10.1039/C4RA13787F

[205] Vander Hoogerstraete, T., Wellens, S., Verachtert, K., Binnemans, K., Green Chemistry, 15[4] 2013, 919-927. https://doi.org/10.1039/c3gc40198g

[206] Vander Hoogerstraete, T., Binnemans, K., Green Chemistry, 16[3] 2014, 1594-1606. https://doi.org/10.1039/C3GC41577E

[207] Sun, P., Huang, K., Song, W., Gao, Z., Liu, H., Industrial and Engineering Chemistry Research, 57[49] 2018, 16934-16943. https://doi.org/10.1021/acs.iecr.8b04549

[208] Riaño, S., Petranikova, M., Onghena, B., Vander Hoogerstraete, T., Banerjee, D., Foreman, M.R.S., Ekberg, C., Binnemans, K., RSC Advances, 7[51] 2017, 32100-32113. https://doi.org/10.1039/C7RA06540J

[209] Rasheed, M.Z., Nam, S.W., Lee, S.H., Park, S.M., Cho, J.Y., Kim, T.S., Archives of Metallurgy and Materials, 66[4] 2021, 1001-1005.

[210] Nam, S.W., Park, S.M., Kim, D.H., Kim, T.S., Metals and Materials International, 27[3] 2021, 538-544. https://doi.org/10.1007/s12540-019-00605-8

[211] Uda, T., Hirasawa, M., Yazawa International Symposium: Metallurgical and Materials Processing: Principles and Techologies; Aqueous and Electrochemical Processing, 3, 2003, 373-385.

[212] Lorenz, T., Bertau, M., Journal of Cleaner Production, 215, 2019, 131-143. https://doi.org/10.1016/j.jclepro.2019.01.051

[213] Lorenz, T., Fröhlich, P., Bertau, M., Chemie-Ingenieur-Technik, 89[9] 2017, 1210-1219. https://doi.org/10.1002/cite.201600171

[214] Lorenz, T., Bertau, M., Journal of Cleaner Production, 246, 2020, 118980. https://doi.org/10.1016/j.jclepro.2019.118980

[215] Wada, H., Arai, M., Ogawa, K., Yamaguchi, K., Journal of the Japan Institute of Metals, 85[9] 2021, 359-365. https://doi.org/10.2320/jinstmet.J2020062

[216] Takeda, O., Nakano, K., Sato, Y., Materials Transactions, 55[2] 2014, 334-341. https://doi.org/10.2320/matertrans.M-M2013836

[217] Liu, M., Cui, H., Li, Q., Zhu, P., Liu, W., Lu, Q., Zhang, D., Pang, Z., Yu, X., Yu, C., Zha, S., Liu, Y., Yi, X., Yue, M., Journal of Rare Earths, 39[11] 2021, 1396-1401. https://doi.org/10.1016/j.jre.2021.03.014

[218] Hammache, Z., Bensaadi, S., Berbar, Y., Audebrand, N., Szymczyk, A., Amara, M., Separation and Purification Technology, 254, 2021, 117641. https://doi.org/10.1016/j.seppur.2020.117641

[219] Wang, B., Zhang, Y., Sun, W., Zheng, X., Li, Z., New Journal of Chemistry, 45[34] 2021, 15629-15636. https://doi.org/10.1039/D1NJ01958A

[220] Wamea, P., Pitcher, M.L., Muthami, J., Sheikhi, A., Chemical Engineering Journal, 428, 2022, 131086. https://doi.org/10.1016/j.cej.2021.131086

[221] Mir, S., Dhawan, N., Mining, Metallurgy and Exploration, 38[5] 2021, 2117-2131. https://doi.org/10.1007/s42461-021-00446-3

[222] Kim, Y., Seo, H., Roh, Y., Minerals, 8[1] 2018, 8. https://doi.org/10.3390/min8010008

[223] Wansi, E., D'Ans, P., Gonda, L., Segato, T., Degrez, M., Procedia CIRP, 69, 2018, 974-979. https://doi.org/10.1016/j.procir.2017.11.011

[224] Inghels, D., Bahlmann, M.D., Resources, Conservation and Recycling, 164, 3032, 105178. https://doi.org/10.1016/j.resconrec.2020.105178

[225] Namburi, D.K., Singh, K., Huang, K.Y., Neelakantan, S., Durrell, J.H., Cardwell, D.A., Journal of the European Ceramic Society, 41[6] 2021, 3480-3492. https://doi.org/10.1016/j.jeurceramsoc.2021.01.009

[226] Marra, A., Cesaro, A., Belgiorno, V., Environmental Science and Pollution Research, 26[19] 2019, 19897-19905. https://doi.org/10.1007/s11356-019-05406-5

[227] Sethurajan, M., van Hullebusch, E.D., Fontana, D., Akcil, A., Deveci, H., Batinic, B., Leal, J.P., Gasche, T.A., Ali Kucuker, M., Kuchta, K., Neto, I.F.F., Soares, H.M.V.M., Chmielarz, A., Critical Reviews in Environmental Science and Technology, 49[3] 2019, 212-275. https://doi.org/10.1080/10643389.2018.1540760

[228] Innocenzi, V., De Michelis, I., Kopacek, B., Vegliò, F., Waste Management, 34[7] 2014, 1237-1250. https://doi.org/10.1016/j.wasman.2014.02.010

[229] Innocenzi, V., De Michelis, I., Vegliò, F., Journal of the Taiwan Institute of Chemical Engineers, 80, 2017, 769-778. https://doi.org/10.1016/j.jtice.2017.09.019

[230] Talebi, A., Marra, A., Cesaro, A., Belgiorno, V., Norli, I., Global Nest Journal, 20[4] 2018, 719-724. https://doi.org/10.30955/gnj.002646

[231] Li, F., Wang, Y., Su, X., Sun, X., Journal of Cleaner Production, 228, 2019, 692-702. https://doi.org/10.1016/j.jclepro.2019.04.318

[232] Li, C., Zhuang, Z., Huang, F., Wu, Z., Hong, Y., Lin, Z., ACS Applied Materials and Interfaces, 5[19] 2013, 9719-9725. https://doi.org/10.1021/am4027967

[233] Saito, S., Ohno, O., Igarashi, S., Kato, T., Yamaguchi, H., Metals, 5[3] 2015, 1543-1552. https://doi.org/10.3390/met5031543

[234] Peelman, S., Kooijman, D., Sietsma, J., Yang, Y., Journal of Sustainable Metallurgy, 4[3] 2018, 367-377. https://doi.org/10.1007/s40831-018-0178-0

[235] Yang, F., Yan, B., Quan, S., Li, N., Zhang, L., Chinese Journal of Environmental Engineering, 10[4] 2016, 1789-1793.

[236] Yin, X., Tian, X., Wu, Y., Zhang, Q., Wang, W., Li, B., Gong, Y., Zuo, T., Journal of Cleaner Production, 205, 2018, 58-66. https://doi.org/10.1016/j.jclepro.2018.09.055

[237] Khanna, R., Ellamparuthy, G., Cayumil, R., Mishra, S.K., Mukherjee, P.S., Waste Management, 78, 2018, 602-610. https://doi.org/10.1016/j.wasman.2018.06.041

[238] Turner, A., Scott, J.W., Green, L.A., Science of the Total Environment, 774, 2021, 145405. https://doi.org/10.1016/j.scitotenv.2021.145405

[239] Scharf, C., Ditze, A., Hydrometallurgy, 157, 2015, 140-148.
https://doi.org/10.1016/j.hydromet.2015.08.006

[240] Park, S., Kim, M., Lim, Y., Yu, J., Chen, S., Woo, S.W., Yoon, S., Bae, S., Kim, H.S., Journal of Hazardous Materials, 402, 2021, 123760. https://doi.org/10.1016/j.jhazmat.2020.123760

[241] Stoy, L., Diaz, V., Huang, C.H., Environmental Science and Technology, 55[13] 2021, 9209-9220. https://doi.org/10.1021/acs.est.1c00630

[242] Predeanu, G., Slăvescu, V., Bălănescu, M., Dorina Mihalache, R., Mihaly, M., Marin, A.C., Meghea, A., Valentim, B., Guedes, A., Abagiu, A.T., Popescu, L.G., Manea-Saghin, A.M., Vasile, B.Ş., Drăgoescu, M.F., Minerals Engineering, 170, 2021, 107055. https://doi.org/10.1016/j.mineng.2021.107055

[243] Fritz, A.G., Tarka, T.J., Mauter, M.S., ACS Sustainable Chemistry and Engineering, 9[28] 2021, 9308-9316. https://doi.org/10.1021/acssuschemeng.1c02069

[244] Avarmaa, K., Yliaho, S., Taskinen, P., Waste Management, 71, 2018, 400-410. https://doi.org/10.1016/j.wasman.2017.09.037

[245] Tian, L., Chen, L., Gong, A., Wu, X., Cao, C., Xu, Z., Minerals Engineering, 170, 2021, 106965. https://doi.org/10.1016/j.mineng.2021.106965

[246] Chen, L., Xu, J., Yu, X., Tian, L., Wang, R., Xu, Z., Frontiers in Chemistry, 9, 2021, 574722. https://doi.org/10.3389/fchem.2021.574722

[247] Wang, J., Hu, H., Minerals Engineering, 160, 2021, 106711. https://doi.org/10.1016/j.mineng.2020.106711

[248] Hassankhani-Majd, Z., Anbia, M., Resources, Conservation and Recycling, 169, 2021, 105547. https://doi.org/10.1016/j.resconrec.2021.105547

[249] Nie, D., Zhang, Y., Xue, A., Zhu, M., Cao, J., Journal of Rare Earths, 36[11] 2018, 1205-1211. https://doi.org/10.1016/j.jre.2018.06.001

[250] Zinoveev, D., Pasechnik, L., Fedotov, M., Dyubanov, V., Grudinsky, P., Alpatov, A., Recycling, 6[2] 2021, 38. https://doi.org/10.3390/recycling6020038

[251] Zinoveev, D., Grudinsky, P., Zhiltsova, E., Grigoreva, D., Volkov, A., Dyubanov, V., Petelin, A., Metals, 11[3] 2021, 1-21. https://doi.org/10.3390/met11030469

[252] Jalali, J., Lebeau, T., Frontiers in Environmental Science, 9, 2021, 688430. https://doi.org/10.3389/fenvs.2021.688430

[253] Beiyuan, J., Tsang, D.C.W., Valix, M., Zhang, W., Yang, X., Ok, Y.S., Li, X.D., Chemosphere, 166, 2017, 489-496. https://doi.org/10.1016/j.chemosphere.2016.09.110

www.ingramcontent.com/pod-product-compliance
Lightning Source LLC
Chambersburg PA
CBHW071708210326
41597CB00017B/2384